OLD ISTANBUL
& OTHER ESSAYS

About the Author

Gerard McCarthy (1949 – 2022) was born and reared in Dublin and worked much of his adult life as a social worker in Sligo. He studied Philosophy at University College Dublin, with an early interest in Nietzsche and Marcus Aurelius, augmented subsequently by Michel de Montaigne. His first published essays – now collected here – all appeared in eight issues of *Irish Pages* between 2006 and 2021. In his later years, after retirement, he divided time between his Sligo residence, an old schoolhouse on Collanmore Island in Clew Bay, and various travels to the Mediterranean and other peripheries of Europe.

He died in the North West Hospice, Sligo on Friday, 14 January, 2022 in the early evening, a month or so after a diagnosis of cancer. His funeral took place on 17 January at St Colmcille's Church, Rathcormac, Co Sligo, in whose cemetery he is now buried.

Old Istanbul & Other Essays was his first book, a final proof copy of which he received on the afternoon of his death. One of the essays was read at his bedside by his partner Nora in the presence of assembled family members and friends. The Editor's Foreword was subsequently added. It is unclear whether further finished work will be found in the extensive notebooks and diaries he kept over much of his lifetime.

OLD ISTANBUL
& OTHER ESSAYS

GERARD McCARTHY

With an Appreciation by Manus Charleton and Chris Agee

THE IRISH PAGES PRESS
2023

Old Istanbul & Other Essays
is first published in hardback
on 18 January 2023.

The Irish Pages Press
129 Ormeau Road
Belfast BT7 1SH
Ireland

www.irishpages.org

Editor: Chris Agee

Copyright © Gerard McCarthy & The Irish Pages Press

All rights reserved. No part of this book may be reproduced,
stored in a retrieval system, or transmitted in any form,
or by any means, electronic, mechanical, photocopying or otherwise,
without prior written permission from The Irish Pages Press.

Typeset in 14/18 pt Monotype Perpetua
Designed and composed by RV, Belfast. Printed by Bell & Bain, Glasgow.

A CIP catalogue record for this book
is available from The British Library.

Dust jacket photographs courtesy of Gerard McCarthy.

ISBN: 978-1-8382018-0-7

CONTENTS

The Helmsman: Gerard McCarthy (1949-2022)   7

Old Istanbul *(2006)*   23

Old Jerusalem *(2009)*   37

Home from Andalucia *(2011)*   59

The Road to Granada *(2013)*   81

The Silence of Seamus Heaney's Father *(2014)*   103

Mitilini Harbour *(2016)*   113

A Day in Dublin *(2018)*   137

At Algeciras *(2020)*   149

Author's Afterword *(2021)*   169

Acknowledgements   175

THE HELMSMAN: GERARD McCARTHY (1949-2022)

*An Appreciation
by Manus Charleton and Chris Agee*

MANUS CHARLETON

Gerard was a close friend for over 50 years. We met in 1968 in University College Dublin where we attended the same classes in philosophy. From the early 1980s he lived in Sligo with his beloved partner Nora close to where I lived with my wife and children. He had a deep influence on me, and on all who knew him.

Aristotle wrote that one of the many benefits of friendship is that it enables a person "to share with his friends that occupation, whatever it may be, which forms for him the essence and aim of his existence." Gerard understood this as a matter of course. From time to time it featured in his conversation and discussion with close friends. He was interested in what mattered to us most and how we were getting on in shaping it at the heart of our lives.

As a young man in the late 1960s, Gerard and some friends went on camping trips from Dublin to the West of Ireland. It led to their purchase in the early 1970s of a small disused National School on the shore of Collanmore Island in Clew Bay. Gerard went there frequently over the years, often on his own, travelling out and back at the tiller of a small currach that had an outboard engine. Poking fun, we sometimes called him "the helmsman". But there was an appropriate element in that epithet. It stood as a metaphor for the way he perceptively guided us towards a reflective return to ourselves from the midst of busyness and distractions. He kept us in touch with our deeper selves.

Located at the western edge of Europe, the island's isolation and distance from events enabled him to develop a broad perspective on the world as our common home, a home blighted by conflict and displacement. It was a perspective informed by his travels, notably to Istanbul and Jerusalem to experience their heritage of religious cultural differences, and to the island of Lesbos to see the camp for refugees from the Syrian civil war and elsewhere who were seeking asylum in Europe. He was also drawn to India, to Varanasi in particular. The combination of the island and travel enabled him to find a language and style needed for doing justice in his essays to his perspective on the world as our common home.

He was a man for going on solitary walks, usually along a deserted Sligo strand, or in Hazelwood in the early morning. He also initiated a monthly walk with friends which took place every month for years. We often climbed Ben Bulben, or up "Where the wandering water gushes / From the hills above Glencar", as Yeats described it. On these walks the enjoyment of nature was combined with convivial banter, shared reflections on life and events, and companionable silence.

So much of who he was came from the sense he gave of being present in the world – to himself, to others and to nature. He was both singular and sociable as well as empathetic and considerate, traits that inclined him towards his job as a social worker. An older meaning of "considerate" is "showing careful thought", and Gerard always did. He was also stoic. Marcus Aurelius' meditations were a foundational influence. So, too, was Nietzsche's penetrating and disconcerting scepticism. In later years he was also influenced by Montaigne's seminal and effervescent essays.

An annual winter solstice walk with friends included taking a group photograph. Gerard knew the rock on which to place the camera, and the angle at which to point the lens. The resulting picture captured us in another year of aging against an Atlantic backdrop with its tides that wait on no one. Last winter solstice illness prevented him from taking part in the walk. He was on his own last journey, leaving behind essays of great depth and humanity.

Winter 2022
Sligo

CHRIS AGEE

I had the privilege – the noun is correct in this case – of meeting Gerard at length in Dublin on one occasion. We met al fresco at a café across from the new Smock Alley Theatre in a quiet corner of Temple Bar facing the Liffey. Our purpose was to discuss this very book, untitled as yet.

Over the previous decade I had published in *Irish Pages* the first six of the essays collected here, of which one had been placed in the prominent opening position, with the other five appearing close enough to the front, all this constituting in my mind an intended and unique endorsement of a quite unknown writer. Gerard had tentatively agreed to the prospect of a collection; but now we had to discuss timescale and whether further essays might be in the offing before publication over the next year or two. In the event, two further essays arrived and were published in the journal, plus an afterword to what had been titled by then, simply, recording his singular début, *Old Istanbul & Other Essays*.

Gerard was soft-spoken, easy-going, self-effacing, modest, somewhat shy, both deliberate and sparing in his sometimes hesitant, tentative and/or halting words. A very good listener, who tended to respond rather than proffer during the course of our conversation under windy and overcast skies. In hindsight (where one attends, often, more fully to what was fleetingly registered at the time), I can discern quite clearly in him a certain writerly diffidence – not so much about the writing itself, as its appearance in the public sphere after decades as an entirely private activity, in notebooks and diaries.

Like Kafka, a key presence in these essays, he could not have been unaware of his writing's worth and uniqueness – but equally, like that "harbinger of the terrible fate of his people at the hands of the Europeans" who asked his literary executor to destroy his writing (all of it unpublished), Gerard too seems to have felt the pull of a siren-song silence in both his life as lived, and his work as written. The words *silence* and *silent* appear many times in these essays. After all, most major writers must feel that they write, in the first instance, on some level, for themselves – and so it's not necessarily a given that publication should be sought if the essential writerly purpose or uses are otherwise. As appears to be the case with Gerard's writing for much of his life:

> "I remembered the challenging clarity of the question in my own adolescence: how does one live one's life? A question whose answer we can never know beforehand as each of us takes our whole life to answer it. I thought, amidst the cacophony, to listen out for a tone, and to follow it, would be the height one could hope for."

As well as airports, taxis, buses, trains, hotels, mosques, churches and monuments of all sorts, the café is one noticeable *locus classicus* in these essays. Here Gerard steps out of his "wandering", "loitering" and "wondering" (these verbs appear frequently) to take stock of one small corner of humanity at closely-observed and rigorously-empirical quarters, usually over a beer. The café in which we were sitting that autumnal afternoon would have been a world away from equivalent places in the "Old Dublin" of Gerard's formation, such as Bewley's and Burdock's, immigrant Italian cafés, pub grub and instant coffee – that deeply backward-looking Dublin which so struck me on my arrival in Ireland in the late seventies, just four years after accession to the EEC.

Instead, the neo-traditional, twenty-first-century, Continental café model now spread to every corner of the European Union was in full swing around us: big all-purpose espresso machine, croissants and pastries, shelves of select food items, metal tables and chairs, the clatter of baristas, the chatter of customers. (This one, seemingly, owned and run by Romanians.) In retrospect, as with these essays, our in-person hour suggests in its blending of "Old Irish" personal formation and contemporary transnational quest, that Gerard was very much "a stranger at home" in this new European world, whether in Spain or Greece, the West of Ireland or Dublin.

There's not a shred of nostalgia or sentimentality, not a soupçon of the "rare oul' times", in these essays whose "region" (as he describes it) is "where Philosophy merges into Literature, with a preference for a language of metaphor rather than of abstract reasoning." So in our meeting, there was a distinct sense of person and writing being of a single piece; as he says of Kafka, "there is no boundary in any of his writing between the literary and personal".

Likewise – in hindsight again – the district surrounding our café meeting now puts me in mind of the overall European transformation these essays subtly record at points. For many years this short stretch of an older Georgian Dublin along Anna Livia was a still-undeveloped dark oasis twinkling with stained glass and a few lights on the far-side of Temple Bar's rambunctious cultural development. Now the converted theatre, the café, a nearby hotel, spruced-up Georgian buildings, a lawn, tree-plantings and bollards have homesteaded the past with a new cultural space.

One filled too with nearby ghosts: Dublin's Viking origins, Handel's *Messiah*, the needless obliteration of Wood Quay. In other words, the pleasant district where we sat with coffee has become another layer of Janus-faced palimpsest, religious, historical and cultural. And exploring such ancient palimpsests in the two tenses, past and present continuous, in the West of Ireland and other peripheries of Europe, is the precise territory which each of these succinct essays traverses with such magisterial philosophical intent.

Along with Kafka, only five other major writers appear in these essays: Marcus Aurelius, Seamus Heaney, Patrick Pearse, W.B. Yeats and Mohamed Asad (né Leopold Weiss). It's a curious pantheon, and not just at first glance. Gerard was surely deeply versed in the overall oeuvre of each, but their appearance in these essays is stylistically confined to the profound philosophical importance for him of certain of their passages or metaphors. These quintessential allusions appear in suddenly revelatory passages, clearly key touchstones tested over a lifetime of philosophical reflection and (one imagines, for him) advances not only in actual philosophy but perhaps in human consciousness itself.

Taken together, cumulatively, like the careful assembling of a mosaic, these proxy allusions constitute part of Gerard's special brand of philosophizing, one privileging "a language of metaphor rather than of abstract reasoning". In contrast to these literary allusions, Gerard himself is much more reticent, implicit and tentative when it comes to direct philosophical speech, preferring more oblique observations like "I listened to the silence"; or "yes I am drawn to the mosque, but only outside it, in its vicinity"; or "one star floating in the profound silence"; or "in the dying light an atmosphere, a mood, a tone so intimate yet so remote there was no word I could find to reach it."

So the reader is obliged to combine his proxy mosaic of literary allusion with the beauties of his exquisite prose so as to see in the round his attempt to gain a measure of "my own latest small journey in my unknown, my opaque ambition". The result is a bijou masterpiece of distilled literary philosophy, surrounded and infused with marvellous writing on contemporary culture, the great religions and their clashing entanglements, contemporary personal quest and the human tragicomedy itself.

Seamus Heaney seems to me the pre-eminent substantive influence among those listed above, with the others deployed more as exemplary guides along "the road" of the life-journey, towards the reflective and questioned/questioning life. Although a decade younger than the poet, and whether or not he was aware of this quote, Gerard and his work surely belong to the same "moment of transition" described by Heaney in speaking of the generational-historical frame of his own work:

> "I believe the condition into which I was born and into which my generation in Ireland was born involved the moment of transition from sacred to profane ... the transition from a condition where your space, the space of the world, had a determined meaning and a sacred possibility, to a condition where space was a neuter geometrical disposition without any emotional or inherited meaning." (1988)

Gerard quotes and then cites again a Heaney passage from *Stepping Stones* highly cognate with the one above: "there has to be something more than neuter absence". Before himself reflecting towards the end of the book, in a brilliant twist of synthesis and (for me) a true advance in consciousness: "I thought: one couldn't say the universe is without meaning as a microcosm of it, the human being, is so powerfully deeply drawn to it." Which is precisely what *Old Istanbul & Other Essays* so beautifully demonstrates, via his own unique contemporary quest for meanings.

Apart from myself, readers of *Irish Pages* and a few literary friends, Gerard was during his life completely outwith the Irish writing scene, a total "literary nobody". Even some of his siblings were unaware of his writings (published and unpublished), so private an activity it was, often done at the Old School House on Collanmore Island, Co Mayo, in Clew Bay, a photograph of which appears on the cover of this book.

So with Gerard …

No creative writing courses. No "mentoring". No rackety prize competitions. No know-nothing programming deploying "the hearsay heresy". No editors glorying in the rushed role of literary gatekeeper, or upholding the herded tastes of an era. No identity purveyors. No reviewers, agents, publicists or marketeers. No readings, residencies or bursaries. No pitches. No nothing, in fact …

Except the work.

Yes, it is still quite possible. All that really remains needed are paper and pen (or computer), experience and its knowledges, intellect and real "education", reading and reflection, and a good inner critic. So that — as the proverb from the Irish has it — *time and patience bring the snail to Jerusalem* ("am agus aimsir bhéarfaidh sé an seilide go hIarúsailéim")…

In the end, in my opinion, this is the essence of most of the greatest writing.

Nothing proves it better than *Old Istanbul & Other Essays*.

———

Gerard McCarthy died in the North West Hospice, Sligo on Friday 14 January 2022 in the early evening, a month or so after a diagnosis of cancer. His funeral took place on 17 January at St Colmcille's Church, Rathcormac, Co Sligo, in whose cemetery he is now buried.

After an unfortunate printer's delay, the final bound-proof copy of this brilliant and profound book was placed in his hands the afternoon before his death — and its final essay, "At Algeciras", with its perfect last line at this last moment, was read aloud by his partner Nora to family and friends assembled at his bedside as he lay medicated the next day.

For now, there appear to be no other extant, completed essays. But there will be searches.

Gerard was a great and unforgettable person, and a great and unforgettable writer. It is an enormous personal and familial tragedy that his life was cut short at such an age, and that many of his future readers will not have the opportunity to meet the living person that shines through the writing left behind on his "final journey to the final boundary, from which no voyager returns."

Autumn 2022
Žrnovo, Croatia

For Nora
&
For Brenda and the Garçons

OLD ISTANBUL

I flew there via Rome. To Istanbul, Constantinople, Byzantium. *(What's in a name? Does it matter to a city what one calls it?)* It was night when the plane landed. There was a man outside the customs waiting, from the Ambassador Hotel. Sultanahmet. He had a placard with my name on it and he drove me in. The moon was just past its full. It was the beginning of the second half of Ramadan. The month was November. He said that fasting during Ramadan is not so bad in the winter, that the days are shorter and there is no heat to give a thirst like a summer one. He told me his wife had had their first child a few weeks before, a daughter. As we came in he pointed out the Sultan Ahmet Camii, the one that is called in English, The Blue Mosque. The hotel was close by it. He dropped me there. Much of that night there was spent between waking and sleeping. Flitting dreams of home.

In the morning I went wandering, first to Aya Sofia. It was built in the fifth century as a cathedral, and was the centre of the Christian Byzantine world for a thousand years, until in the fifteenth century it became a mosque when the Muslims took over. It was built at a time when the earth was still seen as at the centre, and the dome was seen, by both Christians and Muslims, as a reflection of the heavens. The transition from cathedral to mosque was easy. The Muslims covered the frescoes inside with plaster, and put four minarets around outside it. In the twentieth century the building was made into a museum by the new secular government. As I walked the balcony I could hear the echoes of many tongues below. Looking down, I could see the crowds from different continents, their heads craning upward to the dome. I wondered, is it only in the anonymous eye of the tourist that religions can be reconciled?

I went from there to Topkapi palace. Palace of the Sultans. The harem. Their private quarters where they went through the rituals devoted to the task of bringing about a stable transfer of power to the next generation. Their duty to produce one single reliable heir was paramount over all human feeling. For example, there was a period when it was the first duty of those who became sultans to kill all their brothers. The palace was a museum except for one room, which held what were said to be hairs from Mohammed's head, dust from his tomb, and other relics. There was a cleric in a cubicle with a book from which he was fervently singing. The place was a shrine. Most of its visitors were pilgrims. It was a relief to breathe their more ethereal air.

The loss to the Muslims when Aya Sofia became a museum was not a great one. They had their Blue Mosque just across from it. The domes and minarets of the two buildings mirror one another. In the evening the Blue Mosque looked festive with its light, and the crowds going in and coming out of it. In the middle of Ramadan it was a living place. I was lured inside. It was filled with worshipers. I was standing near the door when a man came and told me to go inside and directed me saying, *sit there*. But then when he saw me sitting down he returned, gesturing. He had meant I should kneel. I did so, strangely, as a cleric read from The Holy Book to the crowded congregation. I left on the first tide of people leaving. The man was behind a table outside the door. When I gave my small contribution he asked me my name. Then he told me his and we shook hands warmly. Outside in the courtyard I heard music and I followed it. There was a festival associated with Ramadan all along the ancient Roman Hippodrome. After a day's fasting, the festive eating. It took place every evening. There were stalls and cafes and music. It was a carnival atmosphere, unfuelled by alcohol. An infectious good will. I wandered through it, drinking it in.

The next day I took a ferry from the Galata Bridge up the Golden Horn to Eyup. It is one of the most sacred of Muslim shrines, the burial place of Eyup Ensari, the standard bearer of Mohammed. A Muslim shrine in a far corner of Europe. I went in to the tomb and was swept along by the oriental fervour of the crowd as they thronged around the tomb and around a glass that was said to contain the footprint of Mohammed. There was a path behind the shrine, leading up through a graveyard crowded with the tombs of those who wished to be buried beside it. On the way I met a gypsy woman on the steps, selling packets of paper handkerchiefs. It seemed apt to buy one, from a woman selling handkerchiefs in a graveyard. Back down around the shrine the light was fading. Many were queuing at a marquee for food. There was an air of expectancy as they then sat with their covered plates, waiting for the call from the mosque that was the sign that they could begin eating. The ritual of the days of Ramadan determined by the passage of the sun. Its duration by the waxing and waning of the moon. The rhythm of Ramadan. The Muslim way of harbouring themselves in the human world.

My days in Istanbul continued, wandering. I had been there many years before but had only shards of memory of it. Then it had been a crossroads, a stopping-point on a journey eastward to an unknown Asia. Lost, I had hardly seen it. This time it was also a crossroads. But this time I was heading westward, on my way to visit a friend who was working for an international voluntary organisation in Kosovo. I walked to the railway station for the night-train to Sofia. The journey began in confusion, searching for the correct carriage among carriages that were going to different places. It brought me back. I saw a man piling coal into a stove at the end of one of the carriages. He wasn't helpful. But then another kind caretaker-man showed me the way to my compartment and left me bedding for my bunk. I had the compartment to myself and I lay down and dozed between waking and sleeping, lulled, as the train headed out from the capital through western Turkey. At four in the morning, I was woken when it stopped at the border-post with Bulgaria. A dark provincial station. I looked out and saw a line of people on the far platform heading towards an office to have their passports checked. The scene seemed haunted by ghosts from the Cold War. I found myself following them through a tunnel to the far side, to the office to have my passport scrutinised. The scrutiny of the border-guard. Looking from photograph to person. *Is this the person I see before me?* When I returned to the train I couldn't find my carriage. The dark platform was almost deserted. I saw one woman getting into her carriage. I went up to her and pointed to it, asking her with one word, *Sofia?* But her one word of reply, *Bucharest*, was uttered in a soft sad guttural melody. Never did any place sound so exotic, nor so far from where I was going. I retreated in confusion. But eventually, as these things normally go, I found my compartment and was back in my bunk before the train headed off. Then in no length it stopped again. A loud knock came to the door. A burly border-guard in uniform stood there, his young female assistant beside him. He had an air of superficial cheerfulness when he first looked at my passport, saying *Dublin*, when he saw it. But his good humour seemed to evaporate when I echoed the word back to him as if perhaps he thought I was mimicking him. It seemed to my bewildered eyes as if he began to manhandle my passport with the conviction it was a fake one. I feared that at any moment he might ask me to leave the train with him, but they eventually departed, the door closed again and the train continued its journey, until suddenly it stopped again. For two hours it was still and silent on a desolate siding as the night lightened into a cold dawn. The silence was broken by the voice of a woman in another compartment. Crying. She was speaking what seemed to be a number of different languages including English, complaining about the cold, saying that she had a child, that she had paid her fare. Could somebody not turn on the heating? There was a male voice trying to soothe her. Then the silence returned, and the train finally resumed its journey through Eastern Bulgaria, where it became a provincial one, stopping at every tiny hamlet. A deeply rural landscape that looked as if time had forgotten it. And yet, through the ages it had been the pathway of armies, some marching east, others marching west. The fluctuating tides of the Christian and Islamic empires. The flow of one the ebb of the other. By then the prospect of sleep had disappeared. I turned to the book I had with me. Marcus Aurelius. *The whole universe is change, and life itself is no more than what you deem it.*

My friend met me in Sofia and we took a bus together to Skopje. From there we crossed the border into Kosovo, to Gjilan. On my first day there I read the news in an internet café. Two truck bombs had gone off in Istanbul. They had exploded outside two synagogues, belonging to a community of Jews living in Istanbul whose ancestors found refuge there centuries ago, fleeing from the Spanish Inquisition to the cosmopolitan tolerance of the capital city of the Islamic Empire. That empire in its heyday stretched far into Eastern Europe, and at one point reached the gates of Vienna. In the late fourteenth century the Muslims defeated the Christian Serbs at the battle of Kosovo. The Serbs have a revered cycle of ballads lamenting their defeat there. Centuries later, Kosovo is now ruled by UN mandate. One of the fault-lines. One of the blurred borders. There were international soldiers on guard outside its Christian churches.

After some days there I bade adieu to my friend in Skopje and took an overnight bus back to Istanbul. It left in the early evening and headed east. The Bulgarian border-post on the road did not seem haunted like the other one. The bus was one of three that make the journey every day from Skopje to Istanbul. My fellow passengers had the cut of ordinary people about their business, on their way to the big city, which, after the fall of Communism, has once again begun to radiate westward. We headed on into the heart of Bulgaria. Around midnight we stopped at a restaurant in the middle of nowhere. There was an old woman sitting outside the toilet there, collecting coins from customers. I handed her a scattering of coins that I hoped were Bulgarian. She handed a couple of them back to me, muttering something, as if I had given her too much. The bus then carried on to the Turkish border. The border-guards in their cubicles were busy eating their plates of food before sunrise. On through western Turkey the sky at the eastern horizon was reddening. The final sliver of the Ramadan Moon was rising above us as we sped along eastward, by bus to Byzantium.

From the bus station I got a taxi into Sultanahmet. This time I checked in to the Star Holiday Hotel. A good hotel, with a room into which the sun shone, and a window from which the Blue Mosque was obscured only by trees. Below, the sights and sounds and smells of the streetscape were reaching up to me. I headed out to them. A man in a shop told me that two more bombs had gone off. The British consulate had been hit. The consul was dead. He was by all accounts a cosmopolitan, who loved Istanbul and planned to retire there. Roger Short. His name now belongs there.

I spent another few days wandering around the old city. On my final morning I walked down through the streets behind the Blue Mosque to the sea and walked along it. The sun was hazily shining on the water. Caught in its misty light there were many boats plying the stretch between Europe and Asia. I came back along the Bosphorus past the international press centre. A few days after the bombing it seemed deserted other than a gate-man chatting cheerfully into a phone. Its satellite dishes were silent. On my way back to the hotel I visited the museum of archaeology. It was filled with Egyptian, Greek and Roman antiquities. In one room there were two sculptures of Marcus Aurelius, side by side. As I looked into his face it was almost as if I could hear him saying, *In the universe Europe and Asia are but two small corners, all the oceans' waters no more than a drop.*

I spent much of that evening in the vicinity of the Blue Mosque. I sat for a while among the crowds, at the seats between it and Aya Sofia, remembering sitting there years before, watching a *son et lumière*, a light show with historical commentary, none of which I could remember. The show was stopped for the winter but there was no need of one. The living place provided its own commentary. The mosque was lit brightly, and inside the lights I could see figures in the portals. From a loudspeaker the praying seemed endlessly to continue, while groups of people were drinking tea around me. One young man approached me and asked what did I think of Istanbul. I said my image of Ramadan had been a much more severe one and that I was surprised at the carnival atmosphere. We smoked a cigarette as we talked. When he put his cigarette out on the ground I asked him did he avoid smoking during daylight hours. When he said he did, I said I didn't feel able to do that. He said I could if I was a Muslim. I went inside the mosque. There was a devout congregation of men on the floor, listening to a cleric speaking forcefully, leisurely, with all the time in the world, confident of a submissive audience. It put me in mind of the Irish Catholic Church of my childhood. I didn't linger. Outside in the courtyard, people were coming and going. Some were hurrying in for prayer. I thought, *yes I am drawn to the mosque, but only outside it, in its vicinity.*

I wandered for a last time through the carnival on the Hippodrome. The family crowds were still flowing. The place was alive with voices and music. I stopped at the obelisk of Theodosius. The emperor of that name pillaged it from ancient Egypt in the fourth century. Its still clear-cut hieroglyphics rose into the night sky. There was a popcorn seller, a punching booth, and a carousel filling the space underneath it. I walked on to the Column of Constantine Porphyrogenitus. Its origins are unknown, but it has been called after the otherwise forgotten Christian emperor of that name who refurbished it with gilded bronze plates in the tenth century. In 1204, during the Fourth Crusade, the plates were ripped off by pillaging western Christians, when they abandoned their mission to liberate Jerusalem from the Heathens. They stopped at the crossroads, and instead decided to take Christian Constantinople. The monument now stands opaque, ungarlanded, deserving its other more common name, *The Rough Stone Obelisk*. That evening there was a toy train circling around it. I stood inside the circle, underneath the obelisk. The rough stone rose into the night sky. Sitting in the train with her young child, there was a woman covered in a dark veil. The gleam of her dark eyes was all that was visible as they passed round in one of the tiny carriages. I wandered back among the stalls, and in again to the courtyard of the mosque where I bought small gifts from some of the women selling there. I loitered, feeling unnoticed in the crowds, glad to feel in some small way a part of them.

What do we have to give to the Muslim world? Our contemporary western one that has lost confidence in transcendent value. Our world that has lost a centre. What more could we show them than our open-minded scepticism, our tolerance for the other point of view, founded in the belief that the world is wider than any subjective interpretation of it? Our belief approaching conviction that each perspective is part of the human story, including the Muslim one. And, having lost ours, an appreciation of the single light cast on their behaviour. Their centre of gravity.

I returned to the hotel room and packed for home. Outside the window the streets were still alive. Above the trees the minarets lit up into the darkness. I wondered would it all soon evaporate like a dream? I lay on the bed for a few hours between waking and sleeping. I could hear the singing of birds in the trees, perhaps made feverish by the din of the streets below. The sounds intermingled, reaching in. At one stage I woke in the midst of a dream of home, but of long ago. A suburban house in Dublin. A fond farewell from the family group standing outside it. In the middle of the night I went down to the lobby where a man with a minibus was waiting. I sat beside him as five Turkish people sat in the back. There was silence for most of the way out to the airport. Mine was broken when the man asked me where I was from. I said Ireland. *Irlanda.* Then we lapsed into silence again until we drove up to the airport and he asked me, *International?* I said *yes, international.* He was told to park away from the terminal by the jumpy police who were patrolling outside it.

Next there was the flight home. The bus from Dublin airport into the city was driven by a man from North Africa. I loitered over a drink in the bar of the railway station, waiting. The window beside me gave out into a view of the street. In the distance, at the end of it in O'Connell Street, I could see the gleam of the new millennium spire. A spire without history. Meaning nothing other than itself. It put me in mind of the rough stone obelisk. Obelisk with an opaque history. Human life tumultuously continuing all around it. Next there was the train down to Sligo. As I sheltered from the rain in the doorway of Sligo station, three black people came out past me bearing boxes of food. One woman was carrying a box of frozen fish on the top of her head. The West of Ireland. Far corner of Europe. The waters of the wider world were washing up against it.

OLD JERUSALEM

Dawn was breaking on the road to Jerusalem. The sherut from the airport carried me in, through the new city to the old city walls. It dropped me at the Jaffa Gate. From there it was predictably a short walk in to the Jaffa Gate hostel where I had booked a room. The alley was deserted. The door of the hostel was locked but there was an envelope with my name on it. In the envelope there was a key with which I let myself in. Inside there was silence. I stood, taking it in. The silence was broken by the shuffling of footsteps, followed by the appearance of a man of about my own age, who introduced himself as a fellow guest, a Scotsman. He said the proprietor would be along later, and showed me where I could recline on a couch. He told me he was a born-again Christian. I lay on the couch, in a twilight between sleeping and waking, until eventually the proprietor arrived and I was let into my room.

Outside the room was a small outdoor landing with a prospect of the roofscape of Jerusalem, more modest than the roofscape of the capital of the old empire, Istanbul. As I stood there a church bell rang. Suddenly, a loudspeaker was turned on nearby, and the sound of the church bell was drowned out by the loud call of a muezzin. Soon the voice was joined by voices from other mosques, filling the air above Old Jerusalem.

In the late afternoon I ventured out, through the narrow crowded alleyways of the souk, in the general direction of the Temple Mount, the Haram esh-Sharif: one of too many sacred places in the world that have more than one name. It is the mount where the Jews had their two temples, and which has held the Dome of the Rock, the Muslim heart of the city, since the seventh century after Jesus. Soon, I found myself at one of the gateways to the Western Wall, the Wailing Wall, one of the retaining walls of the second Jewish temple that was destroyed by the Romans a few decades after the death of Jesus. I looked down at the bowed heads of the crowds in the plaza in front of the wall. The golden dome glittered above them. I headed back into the souk in the direction of the dome, but an Arab man stopped me and said the way was closed. He said I should go to the Christian Quarter. I acquiesced to his assumed authority and walked back the short distance to the Church of the Holy Sepulchre.

In Jerusalem, there is a greater concentration of holy ground than anywhere else on planet earth. The Temple Mount was probably a holy place for long before the Jews made it theirs. The Muslim shrine of the Dome of the Rock now stands on the same site, along with the El-Aqsa mosque. It is perhaps fortunate that by definition the crucifixion would have taken place away from the temple mount, and not in competition for that sacred space. But the Church of the Holy Sepulchre itself has such a concentration of holy ground that it has been fought over for centuries by the various Christian sects, centimetre by centimetre. There has been a precarious balance between them since the status quo was meticulously codified in 1757, by the edict of a Muslim Emperor. In contrast to the abstract space of their Jewish and Muslim cousins, the church is filled with icons, idols. Inside it I found myself in a throng that was more pilgrim than tourist. Many were fervently kissing the sacred spots that are more myth than historical reality. Legend implausibly has the church as the site of both the crucifixion of Jesus, and his tomb. I joined crowds filing up the stairs past the place that has been deemed Golgotha. On the stairs there was a cowled figure sitting motionless, turned away from me. I couldn't tell was it human or statue until, as I descended the stairs, I turned back to look and saw the black living face of an Ethiopian.

Downstairs was a slab said to be the slab on which the body of Jesus was laid, which my book said was put there in the eighteenth century. One of the crowd, a schoolgirl, was spreading out a pack of biros onto the stone and rubbing each one on its surface, presumably to bring back to her fellow students as a good luck charm for their exams. At the putative tomb there were crowds queuing. An Orthodox priest was herding them like children, reprimanding any who stood out of line. I retreated from the church, outside into the courtyard beneath a ladder that has remained unmoved in some obscure political balance for centuries. I saw with relief a quiet door, but when I went inside I was followed quickly by a voice behind me of an exotic cleric who emphatically said, "private". He gestured towards the church, indicating that I should go back in there to join my fellow pilgrims. I didn't obey his injunction. I left and wandered back through the alleys. Suddenly, a Muslim man pushing a heavy cart around a corner almost glanced its wheel off me. A very young woman Israeli soldier began berating him. I carried on, lost in the labyrinth, until eventually I found my way back to the openness of the Jaffa Gate. I sat outside the walls for a while, wondering what in god's name had brought me here.

    I had been there before, *en famille*, on a day-trip to Israel from Cyprus, way back in the last millennium.

Our tour-bus had skimmed along a pre-ordained itinerary. We had been shepherded around Jerusalem for a few hours. I had hardly seen it. In the intervening years, my memory had become less memory than imagination. All that remained were shards: for example, the sudden loud call from an unseen mosque at the centre of Christendom that our guide did not seem to hear as she marched us at great speed through the alleyway of the souk that she said was the Via Dolorosa. Periodically, over the years, the shards of memory were augmented by images on television: most were of conflict, of carnage, of lamentations. But these images, as almost invariably with images in the media, seemed to be skimming along the surface of the unknown *what* that is going on: the turbulent passage of the human world whose direction, if any, is perhaps unknowable to our finite minds.

Meanwhile, Jerusalem has continued to be a place of pilgrimage for three religions: a centre for those who have through the generations made the extreme gesture of committing their all to their imagination of a beyond that they see as far more real than the flood of the finite world. For the pilgrim the name of the city is less the name of a physical place than a metaphor whose meaning is beyond us. As for this infidel, ever since the brief shepherded visit, I had had in some place in my mind the intention to return: this time to follow where my own feet took me, to loiter in the hope that I might begin to see.

On old maps the classical Roman world is little more than a coloured margin around the Mediterranean. Jerusalem is at the eastern periphery. In contrast to the world of the Greeks and Romans, although so close to the Mediterranean world, the world of the bible is the desert: a place of religious fervour: the place where the one God was born: the God that was divided into three faces. If the Romans had not maintained a garrison in Jerusalem, this God might have remained a little known god of the desert. As it was, a few centuries after they destroyed the Jews' second temple, the Roman emperor Constantine had converted to Christianity, and a century or so later the religion born in Jerusalem had reached beyond the western periphery of the empire, as far as the holy mountain on the west coast of Ireland that became known as Croagh Patrick. Sixteen centuries later, names from that eastern desert world had gained for an Irish imagination the deep mythic familiarity of names from a child's story-book.

On Friday morning I went to Bethlehem. I took a taxi from underneath the walls at the Jaffa Gate. When I asked my driver was he a Palestinian he said "yes, I am an Arab". He said he was born in a hospital on the Via Dolorosa. On the way I asked him to show me Jerusalem's infamous twenty-first century wall. He went around the long way, along the Israeli side of the wall; his Israeli number plates gave us an easy journey. Then as we came into Bethlehem he stopped for me to take a photograph of the wall from the Palestinian side. The wall had many graffiti including one I was able to translate for him: *tiochfaidh ár lá:* the old catch-cry of the Irish republican movement: our day will come.

In Bethlehem he introduced me to my guide, who brought me through Nativity Square that I had remembered from the previous visit as claustrophobic with pilgrims, tourists and hawkers. This morning there was a subdued quiet. We went into the church. My guide was a quiet Muslim man with a modest dignity. He explained he had been educated by Armenian Christians, and he quickly got permission to bring me down into the grotto where an Armenian service was just ending. The priest indicated for us to go in. The guide and I were the only congregation, and, at the end of the service, to my amazement I found myself kneeling, and paying my respects at the putative spot of the birth of Jesus. I tried to listen to the silence. Afterwards, my guide showed me around the rest of the church, and, as we were leaving, I asked about the siege that had taken place six years before, when Palestinian fighters had retreated inside the church, and had stayed inside with Christian Orthodox clergy, with the Israelis surrounding it. It had been yet another episode of the human drama that has drawn the children of Abraham together in conflict, across the generations. My guide seemed to flinch at the memory of the episode that did much to inaugurate Bethlehem's current misfortunes. I asked him about the man who had been shot a few days into the siege as he crossed the square on his way in to ring the bells in the church as he did every day. My guide said that of course he had known him: Samir Ibrahim Salman: my guide said he was a simple man, too innocent to know that he should not have headed out that morning as he had always done to ring the bells of the Church of the Nativity. By what seemed to me an uncanny coincidence, the day was his anniversary. It was six years to the day from the day at the dawn of the third millennium when a shot rang out in Nativity Square and killed the bell-ringer of Bethlehem.

Back outside in the square, my guide told me that he almost never visited Jerusalem as it had become too difficult. He spoke about the problems of living in Bethlehem: he said that tourism and pilgrimages had been their lifeblood and that few visitors had been coming since the beginning of the century. He told me that many of his associates had emigrated, to Europe and America. He brought me to the edge of the square and showed me the road with the dumbfounding wall snaking across it. Samir Ibrahim's innocent eye would not have comprehended it: the vision of a wall blocking the way from Bethlehem to Jerusalem.

Back in Jerusalem, I paid a visit to the Holocaust museum. At the door of the Hall of Remembrance I was instructed by an attendant to wear a Jewish cap made from light cardboard. At the centre of the hall is a flame, such as one might find at the centre of a religious shrine, but on the floor surrounding it are dark place-names from the centre of Europe. For a few moments I was its sole occupant. I tried to listen to the silence. I thought of Franz Kafka: child of Zion: one of its finest, and most sensitive. Harbinger of the terrible fate of his people at the hands of Europeans. There is no boundary in any of his writing between the literary and the personal. He called himself a Western Jew, saying that meant that not one moment of peace was granted to him. He dreamed at times of emigrating to Palestine, in order that he might live; but, as he wrote in a letter, "the temptation beckons, the absolute impossibility replies." If he had gone he might have lived to see the foundation of the state of Israel. As it was he stayed, and his voice belongs to the soul of Europe.

I came back through the new city. The streets had a Mediterranean European atmosphere that seemed a world away from the world within the old city walls. I walked through a loud busy market. It was near closing time on the eve of the Sabbath. I lingered in the bustle of buying and selling, drinking it in. As I came near the old city walls I passed the King David hotel, and read the plaque outside commemorating the day in 1946 that a bomb was planted by terrorists, targeting the British headquarters there. The sign expressed regret that, as the advance warning had not succeeded, many were unintentionally killed. Two years later the British withdrew from Jerusalem, cutting and running, leaving behind them war.

In the dark of the evening, again I wandered the streets of the souk. Again I made my way in the direction of the Dome of the Rock. Again I was stopped, this time by a boy, and again I was told it was closed. The boy said, "I have roof", indicating he would show me a place from where I could look down on the Wailing Wall. Uneasily I followed him down an alley, but there was no need of uneasiness: he brought me to a platform overlooking the plaza. On the eve of the Sabbath there was a surprisingly festive air. A crowd of young people were dancing and singing in a circle, while above them on either side the two Muslim domes were so close to the wall as to form part of the one scene. Then the call came out from the mosque; and the images of the people, the wall, and the domes above them, seemed as they might to the innocent eye of one such as Samir Ibrahim, as if they were rising towards the one God.

I looked up through the glimmer of the roofscape of Jerusalem into the darkness above us all and could make out just one star floating in the profound silence.

On Saturday morning I walked the old walls of Jerusalem that, unlike the new one, however brutal their past, seem to have mellowed with history. I loitered in the souk, following where my feet took me. In the afternoon I found myself in the Christian quarter. Suddenly I came across a young man carrying a huge cross with a crowd following him. Via Dolorosa: I didn't linger on it long. I carried on into the Muslim quarter as far as the Damascus Gate. Hawkers had spread their wares on the steps outside. I lingered in the bustle of buying and selling, drinking it in. As I came back inside the gate, down the steps, suddenly there was a clattering noise and somebody pushed me from behind: it was a man pushing me out of the way of a cart careering down with a man barely controlling it. I carried on until I found myself again at the entrance to the Western Wall. This time I went in. At the bottom of the steps there was a table with cardboard skull-caps. I put one on my head and went down to the wall and sat in front of it. Around me there were various groups, each pursuing their own ritual. Some were unrolling large scrolls. Some men were sitting alone reading the Holy Book. I went up to the wall and touched it for a few moments to pay my respects, filled with incoherent thought. I retreated to the back plaza, and carried on sitting, waiting to see if I would hear the call from the mosque. A flock of swallows suddenly appeared and circled above the worshippers at the wall, but the call didn't come. I went up the stairs to the exit. Then, as I looked down from above, the call came. I gradually returned down the steps; gradually the sound from the mosque receded. By the time I was at the bottom again I couldn't hear it.

I left the Jewish worshippers behind me and within minutes I was once again among my fellow Christians in the Church of the Holy Sepulchre. It had an evening quiet. There was a singing service going on at the putative Golgotha. Some people were making noise on the stairs. A cowled nun at the back turned round and hushed them, but then smiled a sensuous angelic smile to soften it. The putative tomb was quiet. At the Coptic side there was almost nobody. There was a young priest tending it, who handed over to a woman who began cleaning it. She scraped wax from the floor before diligently sweeping it. The priest then took away the day's takings. I thought they were closing up for the night but I saw him nearby afterwards talking into a mobile phone.

The Haram Esh-Sharif was closed to infidels on Friday and Saturday. On Sunday morning I was up early to go to it, but there was already a long narrow queue there when I arrived, just to the side of the main entrance to the Wailing Wall. A young woman from Germany was beside me, who told me that on a visit to Ireland she had climbed Croagh Patrick in the rain. Meanwhile, the Jerusalem sun was rising, glistening in the golden dome above us. As we climbed the narrow ramp to the gate we could see beneath us Jewish men and women praying separately at the wall. Then a small gate gave us an entrance through it to the Muslim sanctuary: the golden dome and the El-Aqsa mosque across from it. It was surprisingly spacious and quiet. I met the young German again, sitting on a low wall, sheltered by trees. There was a group of American tourists beside her. One of the group was an old woman, who stayed on after the others had left, who had a fit of coughing, and who looked at me disapprovingly before she left to follow her companions. Before the young woman left, on her request, I took a photograph with her camera of her standing alone in front of the Dome of Rock. Then there was just the Dome of the Rock. I tried to listen to its silence.

  The Dome of the Rock: is it Art that saved it? It has stood since it was built in the seventh century at the centre of that volatile place. It could be seen as a cousin of Istanbul's Aya Sofia and was built soon after it. Its architecture is as much Byzantine as Islamic. Like the Aya Sofia it has had both Muslim and Christian periods of its history. Unlike the Aya Sofia, it has emphatically not become a museum. It is believed that it was built directly on the ruins of the Jews' Second Temple, around the rock on the mount where, legend has it, one morning Abraham climbed with his son Isaac to make sacrifice. As they climbed, Isaac asked in all his childhood innocence where the sheep was for the sacrifice, and Abraham answered with the terrible words, "God will look after the sheep."

Jerusalem: the mount of the ruined temple is at the centre of the question of it. That place which for millennia has been seen as the human world's opening to the transcendent. Its subterranean passions: fault-line between Christianity, Judaism and Islam: its history is at the centre of the conflict between them. The place where the One God was born. The God that was divided into three faces. The very personal God of Genesis: a figure of authority and unpredictable power, like Kafka's father. How long before there comes a generation that will not be willing to sacrifice its children?

In the afternoon my feet took me without knowing back to the Damascus Gate. I went outside and walked up the Nablus road. By then the sky that had clouded earlier had begun spitting rain, and there was a strong wind blowing. There was a Palestinian girl walking up in front of me, putting out stickers saying *Jerusalem is my city*. Some of them came unstuck in the wind and billowed off, joining the other detritus blowing down the street. I passed a Palestinian bus station. There was a bus for Ramallah just leaving. There was a feeling of having crossed a boundary to a poorer world. But then, further on, I passed the fortified American consulate, and a short distance beyond that was the American Colony Hotel. The doorman welcomed me in. There was an atmosphere of discrete gentility in the foyer, with just a scattering of people around. In a small lounge beside it there was a western journalist with two well-dressed middle-eastern men. (I couldn't tell their nationality or persuasion.) There was a television camera on a stand beside them waiting, presumably for an interview with some important personage. I found my feet retreating out the door.

I walked back the road I had come. There was little sign of human activity, and it was disconsolate in the wind and drizzling rain. A man appeared, coming up the other way, and he approached me drawing out a concertina of postcards. I could see the gleam of the golden dome among them. So used had I become to such approaches I said "no thanks" and walked on. But then, there was something about the man's immediate acceptance of my refusal – a thin poor man with his goods in a black plastic bag, perhaps returning home from the old city with the day's meagre takings. I called him back. He said they were five shekels. I had already paid twice that for less. I gave him the money and he reached into his bag and handed me a selection of cards held closed with an elastic band. Then, as he was turning away, looking down, he said, haltingly: "because – you know – they are for Old Jerusalem." I put the cards into my bag and as I walked back down the windswept road to the Damascus gate, suddenly, without warning, for the first time since I came ... what am I saying? For the first time in a long time ... I wept.

That evening, I paid my final visit to the Church of the Holy Sepulchre. Upstairs at the putative Golgotha there was a Greek Orthodox service. There was then a procession downstairs, and my way was blocked by one among them, an acolyte, only a boy, as they performed their ritual at the putative stone. Then they carried on to the putative tomb. There was a Catholic service in an oratory close by. I stood between their separate musics. They were not singing from the one hymn-sheet. Then as the Orthodox moved away it seemed as if they raised their voices to drown out the voices of their Catholic counterparts. None of them seemed to be aware of the scene in the gospels on Easter morning: when the women came to the tomb to anoint Jesus and were met by a stranger who asked them why they had come to seek the living among the dead – that the man they were seeking was gone.

I returned for my final night in the hostel. In the room I glanced at the book I had brought with me. Marcus Aurelius: *Take it that you have died today, and that your life's story is ended; and henceforward regard what further time may be given you as an uncovenanted surplus...* Outside the calls from the mosques were beginning again. I stood out on the landing to listen until their concordant cacophony died away. I lingered on awhile, letting the sounds drift up to me: church bells, birdsong, Arabic music coming from a radio, the noise of traffic from outside the Jaffa Gate, a continuous stream of random voices coming up from below.

The next morning, as I was preparing for leaving, I was engaged for the last time by the Scottish man who had greeted me on my arrival. A few times during my stay there he had approached me helpfully, in a Christian way. Each time he had taken the opportunity to express his religious conviction, despite my ambivalent responses. This time he seemed to be making a point of giving me his valediction. "You don't believe in anything, do you?" he began. I mumbled, "Nothing that I could put into words." Then he gave his synopsis of how Jesus had saved him: he said his life had gone astray: he called himself the last card in the pack and told me how he had been laid lower than anyone else he knew. He said, if he could be saved by Jesus anybody could and looked at me intently. As he was showing me out the door we looked at the sky in which the clouds were once again threatening. The previous day's rain had been no more than a drizzling. He said it looked like there'd be a deluge today.

I went back out for the last time to the Jaffa Gate and took a taxi to the bus. The driver was a droll man: when I asked him would it rain he said words to the effect that he didn't make conjectures about matters that are in the hands of The Almighty. When he asked me where I was from, I said Ireland, and then delicately asked him the same. He said that he was an Arab: "You are Irish, I'm an Arab, that's the way it is." We agreed it was a pity about the plight of his native place. I didn't ask him did he call it Israel or Palestine. He said he had heard that Ireland had had its troubles too. I told him about the outbreak of peace there: of the day that came that, by an inexorable law of history, was utterly unlike the day that any of the antagonists had conceived.

In the bus to Tel Aviv I was sitting beside a man who couldn't speak, who for a while engaged me in accentuated communication with his face and hands, but it didn't last long. For the rest of the time I looked out the window as we left the desert hills behind us and made our journey down through the land of Israel to the city that is as yet only lightly weighed down by history: Tel Aviv. It is a twentieth century city, but is an outcrop of the old Arab port of Jaffa which it has now assimilated into a suburb. (After I returned home I saw on television an interview with an Arab man from Jaffa who spoke about the time the Israelis had uprooted his father's orange trees to make way for development. He said his father had said it was like his heart was being ripped from his chest.)

When the bus arrived in Tel Aviv the sun was shining. Outside the bus station I put my nose into the sea breeze and walked down until at last I saw the open blue of the Mediterranean. The waves were rough and the breeze was cooling. I passed a young couple on the sea wall, dressed in Orthodox Jewish clothes. They were sitting beside one another in front of the waves, and the tender image transcended the clothes of the particular faith that they happened to be wearing. I spent the rest of the day ambling slowly along, stopping at various stages, with the sun on my face, breathing it in. The beach was punctuated with signs saying go in peace that could not evoke anything other than heartfelt agreement. The sea did not seek to proselytise me. The breakers just kept rolling in. On my way back I stopped at a beach café for a bottle of beer which I drank as slowly as the sun descended to the western horizon. There was a fusion of east and west in the pop music on the radio. In front of the café people were doing their beach things. Surfers were rising and falling in the waves. I watched one distantly small figure persistently heading out through the waves, before a brief riding, before a fall, before heading back out persistently again into the immensity.

When I walked on I saw a man with long grey hair and beard looking out to sea. Unlike the Western garb of the others around him, he was dressed almost like an Arab, or a follower of some obscure religion. It seemed as if he had come out to pay his respects to the setting sun. I imagined Jesus after his resurrection, leaving the desert at last, making the short journey to the coast, and living out the rest of his time as an uncovenanted surplus, in silence by the sea.

I walked back up to the bus station in the darkening day. The bus headed out into the darkness. When it arrived at the airport I sat for a while outside in the fresh night air, close to where I had sat on the night of my arrival, thinking that there was nothing to prevent me boarding a bus back to the Jaffa Gate, and heading back inside the walls to make another attempt at Old Jerusalem. But that was just a passing idle thought. However unknowingly, by whatever mix of chance and intention, the path followed was the one that was taken. The past is inexorable. My feet made for the departure door. On my way in I was stopped by a young security man who looked doubtfully at me and my bag as he asked me the question which I already knew I would be unable to answer: "what was the purpose of your journey?"

HOME FROM ANDALUCIA

Rain was falling on the road to Cordoba. By the time the bus came in to the station it was teeming. I found a city bus into the centre. I could see nothing out of the windows. When the driver kindly called out to me, "Mesquita", I stepped out into a deluge. There was thunder and lightning beginning. As I ran with my bag I caught a glimpse of the mosque above me but, I had no idea how, I found myself between it and the river on empty ground that seemed like no-man's land and I could see no way out of it. A working man came out of his hut and kindly opened a fence for me, letting me up to an alleyway beside the mosque. As I came in around it the heavens opened: there were floods flowing along the road, off every roof, and out of big drain-holes high up in the wall of the mosque. The atmosphere was apocalyptic, millennial. Eventually I found the hotel I had booked into: the Hotel Los

Patios, across the alleyway, directly opposite the main gate of the mosque, known as the gate of the pardons. By that stage I was drenched. I checked in quickly and was shown to my room where I spread round my wet clothes. When the rain eased, I went across and into the courtyard of the mosque: the Grand Mosque of Cordoba – or at least that's what it was until the thirteenth century when the Christians took over. A few centuries later they planted a cathedral in the middle, with the mosque still encompassing it. I wandered in the shelter of the cloisters around the courtyard, imbibing its balanced stillness. Outside one of the doors I could hear organ music. It was closed for the evening. I looked in through a gap between a screen and a pillar. There was a musty waft of old stones and incense, and the flowing murmur of prayers rising up into the unseen. There were still occasional flashes and peals of thunder and a light rain falling. Back in the hotel, before bed I sat at a small desk next to the one window that looked out on a wall and a tiny courtyard beneath, gathering my thoughts. I had been there before ...

Early in the new millennium, the happenstance of life saw me travelling on an overnight train heading south from Barcelona. I was sharing the compartment with a silent countryman. I was lulled to sleep by the rhythm of the train until the middle of the night when it slowed, and started stopping and starting in the dark countryside. I watched from my berth as the other man sat up and began peering intently out the window. When the train stopped briefly at a small sleeping station, seemingly in the middle of nowhere, he took up his bag and disappeared out into the night. The memory has retained the fevered clarity of a dream, and perhaps it was: in any event the next I knew was when I woke alone in the compartment in the early morning as the train pulled in to Granada.

I made my way straight up to the Alhambra and checked in to the Hotel America which is right in the centre of it. I went wandering. The visitors were only beginning. I had two café solos by the ramparts. At the kiosk there were guards leaning with coffee. The handcuffs hanging from their belts seemed incongruous as they chatted cheerfully with others at the bar. The day's work was only slowly beginning. A man came along sweeping his ritual round. On the far side of the kiosk there was a cat strolling through a scattering of pigeons, ignoring them as they did him. It was a scene that would have been as much at home in the fifteenth century. Soon the scene in its entirety was hidden behind the crowds. I wandered desultorily among them through the palaces; I hardly saw them. The most that remained was the sound of water.

I went outside to the street where people were loitering, some selling. One big gypsy woman sat impassively behind her stall. Then, periodically, she would suddenly come to life and let out a loud volley on the castanets at any tourist approaching. Below the Alhambra, dazed and bewildered from tiredness, I sat on a wall. A woman who was passing, suddenly became hysterical as she talked into her mobile phone. It seemed as if she had just been told of a death. She had three daughters with her, of about ten, and five, and a baby. The older one was old enough to start crying as her mother continued to talk hysterically into the phone. The younger one seemed almost untouched, caught in her childhood world. The baby was sleeping. Then a man came and tried to comfort the woman. He took the children away. The older girl was still crying. The woman was still sobbing into her mobile phone. Later I went in to the Church of Saint Mary, built on the site where the main mosque of the Alhambra had stood. Its baptismal font was a bowl that had been part of an Islamic fountain. Near the front I saw the woman I had seen earlier, sitting with her head bowed towards a sculpture of Christ after he had been scourged at the pillar. The bruisedness that is of the essence of experience. Ecce homo.

I retreated to the window of my room in the Hotel America. The sun was glittering in the roof-top in front of me. Above it, behind it, there were a few tall trees in the blue sky. The breeze in the trees was a monotone in the background. Within it I could hear the sound of distant hammering, of water, and a medley of uncomprehended voices in the courtyard below, their tone unhurried and lilting – a southern lilt. Mingling with them were the sounds of dishes being washed, and in the background of the background, playing low, I only gradually heard some genteel popular music from the middle of the twentieth century. Music that one might wear a suit to dance to. I looked at the main headline of the newspaper that I had bought earlier: "The day that Kabul fell". It was 14 November 2001.

By pure coincidence, later on the street I met friends from home. They gave me a lift the next day to Cordoba. In Cordoba we checked into the Hotel Los Patios. I was drawn with a powerful gravity into the mosque and the cathedral that is contained by it. I visited the mihrab, the one remaining Muslim prayer room. I wandered around the perimeter of the cathedral. Preparations were being made for Mass. Priests were fixing their vestments in the side-chapels. One priest was sitting reading in his confessional, waiting for penitents. Then the organ struck up, and the choir. I stood between the mihrab and the cathedral, listening to the fervour of the Christian praying, imagining myself across the divide, wondering, if Church and Mosque were to hold a joint service of prayer there, might there not be a good harmony between them?

Strolling around the streets, I stopped at the statues of three philosophers: Averroes, Maimonides and the Stoic Seneca: A Muslim, a Jew and an ancient Roman, joined together by the accident of their Cordoban birth. Later, I heard above the noise a call in the darkening. Distant. Modest. Almost timid in its unobtrusiveness. I went in search of it, up one alley and then down another, following the voice until I found it: the call from a living mosque. It was no bigger than a living room. There was a small scattering of footwear outside it but I met only one man. He said wryly that it was not a grand mosque. He said that there were no more than a few Muslims living in Cordoba. He didn't say it was once the centre of a Muslim empire, that at the turn of the first millennium it and Constantinople had been two beating hearts at the periphery of Europe. He didn't ask me to take off my shoes.

2001 wasn't the first year that saw Kabul fall. It first fell to an Islamic conqueror in what was for Europe the ninth century of the Christian era. It is of course in the nature of cities to rise and fall. For example, Granada was the last bastion of Muslim Spain before their surrender to the Catholics on 2 January 1492.

The Muslims and Jews were given a choice to convert or be expelled; it is of course in the nature of history that one people's fall is another people's rising. The keynote of the rise of Catholic Spain was struck seven months later, on 3 August, when Christopher Columbus set out from near there, westward for an unknown continent that he thought was Asia. However, less than a century later, in 1588 a Spanish armada set out to fight the British, but met disaster. Their Catholic faith didn't save them. The storm that wrecked many of them off the west coast of Ireland was seen by Protestant Europe as a divine vindication. A medal was struck in commemoration with the words: "Jehovah blew with his winds and they were scattered."

In an ideal world, cultural boundaries should be permeable membranes through which the human spirit would flow. In the world of ideas this can be seen to have been the case. For example, the ideas of ancient Greece over centuries flowed through Egypt and around, south of the Mediterranean, back through Andalucia into Europe. In the real world, all too often the boundaries have been theatres of war – all too often the armies have fought under the banners of religion.

There is layer upon layer of history in Andalucia. It is the south-western boundary of the European story. Before the Muslims came with their dispensation, the Romans ruled there for centuries, around the time of the birth of Jesus. At least for the affluent Romans, it must have been a languorous sensuous life of plenty in beautiful bountiful surroundings. I imagine they viewed the Christians as the world of liberal agnosticism views Muslim fervour: gazing with bemused distaste at the strange fire of religious passion.

Andalucia was a land that was already old before the Romans came with their dispensation. It was one of the gateways into Europe for a species that gave itself the epithet, *homo sapiens*. (A species unknown by that name to any other species.) From what we know, our ancestors came across from Africa, heading on in past the coast of light, through a land that was not yet Andalucia, northward into a land that was not yet Europa, until generations later some of them reached a much darker sea: the north-west Atlantic periphery. The place the Romans called Hibernia: a winter place.

Home from Andalucia at the onset of winter, the gloom was both internal and external. My lot as a provincial social worker had an uneasy edge to it. Meanwhile, the mood of the quotidian was matched by the deterioration in the weather. The evenings were closing in, and the long hours of darkness were increasingly separated only by short timid days. I wondered ruefully, if homo sapiens had evolved in the west of Ireland would we ever have emerged from the shelter of trees?

On a succession of dark evenings I read through an account of the long succession of wars between the Christians and the Muslims over Andalucia. No doubt each generation was consumed in its role, not seeing itself as a brief chapter in a convoluted history of victories and defeats, of actors succeeding one another on the stage. I got lost trying to follow them. All that was clear was that, whatever subjective worlds the protagonists conjured up about what they were engaged in, their private perhaps unconscious motivation was much more material.

I turned to the philosophers. Averroes and Maimonides were beyond me. I couldn't grasp them. And as for Seneca ... the memory remains of sitting with Seneca in the study in the blue chair by the fire, disgruntlement piled on disgruntlement by the acknowledgement that my much more fortunate fate than so many others made such disgruntlement illegitimate, falling far short of Seneca's ideal; alleviated only slightly by the suspicion that he fell far short also: a man of the world writing letters to the world, dressed in the garb of philosophy, in contrast to the intimate musings of the emperor who came along a hundred years after him – Marcus Aurelius: ruler of the known world, in private discourse with himself. His human presence. Unlike Seneca and too many others who enjoin others as to how they should behave, his admonitions were to himself, out of the better part of himself. For example: "Either you continue living here, to which custom has by now seasoned you, or you remove yourself elsewhere, (which you would do of your own free election), or you die. There is no other option. So put a good face on it."

I went for a walk out in Streedagh. On the way down to the beach I stopped at the memorial to those who drowned when three ships of the Spanish Armada went down off the coast there. We have the account of one of them, Francisco De Cuellar, in a letter to a compatriot. He tells of the turbulent days on Streedagh strand as the waves continued to roll in, indifferent to the human flotsam that they threw up on the shore, to be set upon both by the native Irish and the British. His account is that of a man from the advanced world washed up on the shores of a primitive tribal country. The image he paints is reminiscent of the contemporary western image of Afghanistan.

The curiosity of De Cuellar's account: that which he took for granted, which he did not consciously advert to. His account might be seen as a more reliable one than that of the partisan ones of the English and Irish protagonists, as it is a memorial to the subjective experience. Did he have some instinctive feeling that what he put down might be more illuminating to someone subsequently, that they might see that which he did not see, that they might have a wider view from within which to look at it? He could not have imagined that his letter would join the letters of Seneca and a myriad of others, lodged in the ether of the internet perhaps until the end of the human world.

How does one give an account? One must choose one foothold in the flux; one island; one form that might shed light on it; one shape to it; one story told, in the light of the acceptance that the view from one's world is a finite one, part of whose essence is its washing up against other finite ones – and that one should use the lights of others to interrogate one's own.

I carried on down to the small first beach. The swell had waves discretely breaking. Their sound in the round was ample and fresh-cutting. With the sun shining across them, their curve was a particular strong green like the colour of a precious stone, before they exploded into a cloud of white spray. I climbed up to the headland. The sea beneath was as always like a living thing, turning and tossing with the low momentum of the swell that, having originated at a distance, inhabited the water more intimately than the much more transient waves. Restless against the rocks, it was only hinting at its own power. Enigmatic undercurrents. *Keep chipping away.*

I carried on out along the sea wall. Beyond it, the island of Inishmurray a few miles out in Donegal Bay only occasionally came into my notice. It didn't seek it. Its low unmoving shape was numinous, enclosed, self-contained in its own distance. It was only when I stopped to scrutinise it that I could make out the wall of the ruined monastery, the line of derelict houses, and the school the children attended before the island was abandoned in 1948. Since then it has been left to the stewardship of wild nature, its teeming non-human life secluded, unseen by the human world, except for the huge flock of barnacle geese who winter there each year. Most days they fly in to graze fields on the coast of the mainland and come to our notice by the power of their numbers, the pattern of their flying. As I stood there I saw a small cloud swirling above the island, furling and unfurling. The winter geese – what else can they do but keep returning?

The houses on the island are still raw in their dereliction, but the monastery ruins have mellowed with the centuries and it is hard to imagine it becoming a living monastery once again. The monks came there first in the sixth century, when the Christian ceremonials were not venerably old, but fresh and strange and new and, if Patrick is anything to go by, they were fired by a fundamentalist belief in Christianity, and in the imminent end of the world. This faith must have allowed them place the tortuous drama of human history at a remove. But history continued and is still continuing long after their monastery became ruins. It was deserted around the middle of the second millennium. For a few centuries afterwards it continued to be a place of pilgrimage, but its central icon, a statue of its patron Saint Molaise, is now in the national museum.

Inishmurray: far corner of Europe: one of western culture's outer reaches, where people went with a view of history that thought of itself as transcending it. The deep human instinct for the beyond. How all-enveloping the sea must have been for the monks who lived there. Its constant powerful presence must have been a deep part of their religious conception. Looking out from their island retreat the world must have looked to them to be away in the murky distance. Were they able to feed on transcendence? Given the human nature of homo sapiens, their lives most of the time were probably swept along by the social rituals day by day in repetitive succession, consumed in the small dramas of their own exclusively male mundane quotidian. Only rarely would they have been able to go beyond their tunnels of thought and allow the seascape be in the purity of its manifestation.

The monks of Inishmurray: their faith in the literal meaning of their prayers. How does the human need to transcend translate into a time when any one faith is seen as one among many? As metaphor? There remains a faith in the need to transcend old local perspectives, seeking a perspective from which to view them. A belief that what is most central can be found in the boundaries between them.

How can one assume a neutrality when addressing the cultural air we breathe? How can one see one's heritage unless one steps outside it into that space we share in common with all heritages: the world that none of us can step outside – until the final voyage to the final boundary, from which no voyager returns?

On a clear day the monks would have seen nothing out to the west but vast tracts of sea and sky meeting one another at the horizon. In contrast to the African and Asian boundaries of Europe, the Atlantic boundary was the edge of the known world. When they looked out from their western station they must have thought they were peering into the edge of the beyond; they did not know that they were looking out towards America.

No written record of the monastery survives: there is no record of any of them sailing to Byzantium; there is no record of any of them heading off into the loud crowded world, on pilgrimage to Jerusalem. If they had been shown that great twentieth century image of transcendence: the photograph of the blue planet from outside it, might they have abandoned the world of the island and gone to explore the island of the world?

On the way back I stopped at the remnant of a megalithic tomb at the top of the dunes: an ancient grave of a human being, about whose life one could do little more than imagine that they were buried there with ceremony when they stopped breathing, having lived at a time that we now call pre-history: before Homo Sapiens discovered the miracle of the written word. From there I could see across the small bay to the one surviving wall of the small abbey at Staad. It is probably the place that De Cuellar came across and saw some of his compatriots hanging from its rafters. In front of it I could make out the landing place from where the monks used to set out. It must have been an act of submission, the distant opaque image of the island far out in front of them: an uncertain voyage with no knowledge about when they would be returning. Their varied hearts. Out there on a small craft in a huge sea, their fate in its hands. The power of the deep in which one has no option other than to place one's hope and trust, having embarked on a voyage with no other option than to continue it, no option other than to try to hold a resolute direction. *Keep faith in it.*

In the beginning, the new millennium seemed to hold out the possibility of a new beginning. The closing dates of the old century had seemed heavy with the burden of history. The new dates, beginning with zero zero, seemed like a tabula rasa on which a new era might be written. But soon, Homo Sapiens was behaving as it always had done. The attack on Kabul in November 2001 was in retaliation for an attack on America a couple of months earlier, and the digits of that date, nine eleven, struck the keynote of a war that seemed to have neither direction nor ending. Meanwhile, in the West of Ireland, the days of our lives merged into one another, in contrast to the date one might put at the top of a page. Season continued to follow season as they had always done, with the added poignancy that the more of them one had seen the quicker they seemed to come round. And so it was, when I woke once again in 2010, on 8 December, in the Hotel Los Patios, in the darkness of the early morning, it seemed as if the first decade of the new century had passed like the dew of a dream.

 I went across to the mosque/cathedral just as it was opening. I wandered through the pillars around the perimeter of the cathedral. Red velvet ropes were being hung, marking it off from the mosque surrounding it. Inside, in front of the altar there was a float with flowers and a statue of the Virgin Mary, and men in quasi-military ceremonial uniforms. It was the feast of the Immaculate Conception. It took me a while to find the mihrab. I stood in front of it and noted the small area of it and its adjoining spaces that seemed untouched, without icons, while beyond them were altars with florid images. I continued my circling, each time using the mihrab as a marker, often becoming lost before I found it again. Because of the feast-day the place remained open only an hour for secular tourists. Soon the marshalls began herding us. I made one last attempt at approaching the mihrab but a marshall gestured me away from it. Then I saw him unlock the gate. He went inside and turned out the light. Other lights were being turned off. Soon the perimeter of the mosque receded into darkness and only the lights of the cathedral remained on as the last of the secular visitors were herded out the door.

Outside in the courtyard I stood at the pool and fountain. An assembly of dignitaries was gathering at the main door. There were a few smooth-faced men with the cosmopolitan ambience of euro-politicians, but most were quasi-military figures, one with a strange flamboyant hat that made him look like a character from some comic opera. There was much hand-shaking and gentle jovial slappings on backs and shoulders. There were rapid kisses on the cheeks exchanged with the few females. I looked in beyond them into the mosque but could see nothing in the gloom inside. There was a guard of honour at the door holding flags aloft as the gathering filed in. Then they lowered them and followed the others, and the big doors were shut behind them. It seemed that whatever might be going on inside was much more than mere religion.

    I went back out through the gate of pardons to the streets of this world. I crossed the Roman bridge behind the mosque. The river was rushing flooded and brown underneath it, carrying branches of trees unruly in the rough flow. At the far side I went in to the Callahora Tower where there is a museum run by a foundation for dialogue among cultures. I asked the woman behind the desk was there any living mosque now in Cordoba. She said there was just one. Her finger pointed on my map to the Plaza de Colon. The streets were quiet as I walked to the plaza. Most establishments had closed for their siesta. In the plaza there was a park with a fountain in the middle. By now the sun was shining. I sat on a bench. Gradually my eyes were taking it in, until the image of a small mosque at the other side of the park came into my view. A man with a lightly-veiled woman and child came along. He was walking in front of the other two. He disappeared inside the mosque. The woman accompanied the child to a nearby play area and stood by as he played on a swing. I went over to the mosque. It was very small, but not the one I remembered from the last time. There was a young man in the hallway washing himself as a child in a pram kept up a babble that sounded to my ears like "Allah, Allah". I passed in by them. This time I took off my shoes.

I came back to the Grand Mosque and loitered in the courtyard beside the pool where the believers used to wash themselves before going in to pray to their God beyond images. Their words. Their humble words. I saw a man in an alcove, slightly bowed in prayer. Something about his stillness made me wonder might he be a Muslim, but when I turned to look towards where he was praying I saw a crucifix. I headed back out on the Roman bridge. The sun was sinking behind clouds to the west. The mosque was in the background of the swollen river. I loitered among the crowds sauntering across the bridge, enjoying the balmy national holiday. There were a few musicians playing and the atmosphere was appealing. Then the sun disappeared behind the clouds, a cool wind got up, and a few spots of rain were falling. The crowds thinned. There were rumbles of thunder and flashes over the mosque. I thought: the cathedral draws the faithful, but it is the mosque that draws visitors from afar. I remembered the museum of the Aya Sofia in Istanbul, once the centre of Christian, then of Muslim fervour. I wondered: is it only when all places of worship have been turned into museums that religions will be reconciled?

The next morning I was up again before the darkness had receded. I stood across from the gates waiting for them to be opened. A few had gathered with me by the time the bolt was pulled back and a man inside swung them open. Inside the mosque there was a red velvet rope marking off the area of the cathedral. There were cleaners at work, and attendants loitering in a purposeful way. I roamed the aisles, circling the red rope marking the cathedral. A cleric passed me with a briefcase that looked as if it contained a laptop. I heard a few strains of the organ. There was an atmosphere of preparation. Suddenly, as if by magic, a large group of colourfully dressed clerics appeared and a ceremony began. The small congregation consisted mainly of elderly women. No secular tourists were allowed inside the velvet boundary. Then there was singing, and prayers were intoned. I noticed a picture right beside the altar of a military man on a horse seemingly crushing another man underfoot. Once again as I had done nine years before, I stood between the mihrab and the cathedral. Midway between them there was a statue of a saint: a young woman with arms outstretched, eyes lifted upward piously, a sword sticking into her breast. There was a bench immediately in front of her. I sat there listening to the Christian ministers singing to the glory of their one true God in heaven, and looking at the tourists who clustered in succession around the mihrab, taking photographs of it and of one another. The mihrab of the Grand Mosque of Cordoba: is it art that saved it? It has at least now transcended the conscious purpose of its creators and has achieved the status of just being there, beyond any claim about the statement it might be making.

In Malaga on my way home I checked into the Hotel Sur. In the late afternoon I walked down to the beach. Out in the sea there was a distant shimmering that I imagined was Africa. I imagined our ancestors coming across the narrow strait in their fragile crafts. Homo Sapiens: the animal that found itself on a voyage with no option other than to continue it. Some surfers came down and headed into the water. Most seemed young and inexperienced. They spent long whiles biding time, waiting, rising and falling, slowly making their way out against the waves, often disappearing in the foam. But then sometimes persistence was rewarded with a riding: however brief, an exuberant riding. I remembered the beach at Telaviv. I remembered Jerusalem. That narrow opening to the transcendent: the fervour of the Dome of the Rock, of the Church of the Holy Sepulchre, and the Jewish dream of a third temple on the sacred mount. Might the people of Israel not better fulfil their promise if they turned their backs on the Wailing Wall, and began to build on the shores of the Mediterranean a temple without history, a manifestation rather than a statement – like the sea?

Back home again in the West of Ireland I took a walk along the Yellow Strand. The sky was covered with a blanket of cloud but there was a narrow strip of blue just above the horizon. As I walked the sun descended to it and lit the water and the whole strand. The word that came into my mind for it was terrific. Perhaps I should have thought: prodigious, astonishing, awesome. On my way back the sun disappeared and the ephemeral solstice light was fading. A light wind was blowing in from the south-west. The dune grass was swaying. The sound of the surf in the chamber-box of the dunes was more than just a sound. In the midst of it I began to hear a low monotone that I thought at first might be the echoing in the dunes, but suddenly they were a powerful presence: thousands of geese in their patterned formation. I couldn't tell was it their voices I had heard or the low drumroll of thousands of wings. For moments they were overhead and their inhuman crying reached a crescendo, until they disappeared around the hill of Knocklane on their way out to their island retreat. Suddenly all I had was the sea's eternal hshh, and in the dying light an atmosphere, a mood, a tone, so intimate yet so remote there was no word I could find to reach it.

THE ROAD TO GRANADA

At Knock it was a crystal-clear cold early-April morning. Through the window of the plane I saw a man on the platform of a low-loader fling a bucket of pale red liquid onto the wing and with an implement one might use to clean a big window he began to spread it vigorously; they were de-icing the plane; it was four degrees below zero. The plane rose into the clear blue sky, passing over a clutch of cattle in a boggy field. None of them looked up, intent as they were on the meagre grass beneath them. Soon the sky clouded beneath us and we could have been anywhere or nowhere, until we came down through the clouds a few hours later, into Malaga. Outside the airport I boarded a bus which left almost immediately and brought me in through intermittent drizzle, via the bus station, onto the main street where I got off and walked round the corner in the confidence that I would see once again the familiar sign: Hotel Sur.

After checking in, my feet took me with an inevitability down the road to the sea. At the back of the beach I passed a woman who may have been North-African, sitting looking out. I wondered was one of her kin making the short journey from their native home, across the short stretch of water in front of her. As I walked along, the sky gradually cleared and the sun came out. I sat on a spur of rocks for a long while listening to and looking at the water rising and falling in front of me, trying to gain a measure of my own latest small journey into my unknown, my opaque ambition. On my way back along the quay I passed a Buddhist monk from India slowly walking, dwelling in his own thoughts, fingering his beads. Later, as I wandered the lively, voluble festive evening streets of Malaga, I passed an African man kneeling, his arms held out holding a large placard in front of his chest, repetitively intoning his plea like a prayer. I stood at some distance watching him before returning and throwing a few small coins into his plastic cup.

The next morning, I checked out of the hotel and walked to the bus station. I had a coffee in the cafeteria while I was waiting. There were two young North-African men at another table, who seemed as if their wait was far longer than mine. A very tall, thin man came to them and spoke to them as if he had some kind of authority over them. I noticed his very long thin hands.

An old couple shuffled in slowly, and slowly settled in to their places. I think the old man was slightly deaf, and the only feature that was not old in the image of them was the woman's eyes that had the enthusiasm of a child as she leaned towards him speaking to him. I saw them outside afterwards on a bench looking out for their bus, as I loitered gawking around me, until my time came and the old couple disappeared from my life as I joined a crowd of people at one of the bays. Two buses came in. The crowd was clustering around as the driver of the bus at my bay was barking staccatos of instructions, directing some into the bus and others around behind it. His staccato to me was accompanied by a gesture to go around, and when I showed my ticket to the driver of the other bus he nodded quickly for me to enter it. I climbed aboard without checking the destination on the front of the bus. Inside, my absurd shyness had me not seeking confirmation from another passenger, and for the beginning of the journey I had a slight unease that I might not after all be on the right bus, but as we climbed into the mountains, the signs for other destinations fell away and it became clear that it was as was meant to be, and that I was indeed on the road to Granada.

In Granada, outside the bus station I boarded a city bus in to the city centre. I got off at the cathedral and walked around it. Very quickly I was in the Plaza

Bib-Rambla, and had found the Hotel los Tilos. My room was on the fourth floor. The window overlooked a narrow enclosed space, but down the corridor was a doorway onto a balcony which opened out to the city: the square beneath, the bulk of the cathedral behind it, and the hill of the Alhambra rising above us all. I went out and first explored the square. It was crowded. Among the many traders and hawkers I watched a small human figure dressed as Mickey Mouse, selling balloons that were aloft, floating at the ends of their strings. I gave a coin to a man with one arm who was playing haunting music on a harmonium. He immediately lurched forward, I don't know was it in a gesture of thanks or to see what coin I had thrown into his box.

Soon my feet brought me up the road to the Alhambra. I made my way up to the main entrance and purchased a ticket for the Generalife Gardens. In the gardens, the atmosphere was powerfully redolent of early spring, with fresh budding and flowering. The air was crisp and cool, but the sun was warm. I loitered. I took in the prospect of the palaces below that one doesn't get filing through the crowds inside them. There were crowds in the gardens, but it was possible to find more secluded places, in one of which I saw a veiled Arab woman who looked at first as if she was taking a photograph of herself with her phone, but then I realised she was taking one of her children who had been out of my line of sight in front of her. I passed a younger man and an older man who may have been his father. The older one was determinedly choosing his steps with the younger one helping him. The younger man helped the older one sit down on a bench and left him there. The older man was a significant presence on the bench in his dark glasses, holding his stick in front of him, seeming to be taking it all in. Nearby on another bench was another young man reading a book who decades ago might have been me.

I went down to the Alhambra, and into the Hotel America where I sat at a small table beside a sign saying *America es un continente*, and had beer, and a tomato salad and bread that tasted gourmet. There were four men at a table close by. An older man was showing a younger man photographs on the back of his camera. I heard the younger man say, "I thought she was your daughter", to which the older man averred the opposite as he continued to range through the photographs, to the other man's polite dismay. I thought: far better to be sitting alone at a table than to be in company one didn't choose, watching a succession of images of people one doesn't know, interminably. I left the scene behind me and went outside to the ramparts where I stood looking down at the city. It was a commanding prospect. Back down, when I reached the Plaza Nueva, I heard the sound of a marching band, then saw a procession, which included a big float of Christ carrying his cross, as if Easter hadn't yet happened. I headed up an alleyway that had become established as an Arab souk. My feet quickly brought me to a small square with a church at the edge of it: The Iglesia de San Gregorio Betico. I entered its shadowy light. At the bottom of the church there was a sign in a number of languages which said that God was present there, enjoining us to adore him. At the top, behind a grill from floor to ceiling, kneeling in adoration before the Host displayed in the richly ornate altar, there was the figure of a nun, all in white. All in white, she was covered completely; but for her occasional small movements it was possible to doubt that she was a real person. There was a clock on the wall beside her; its loud ticking was occasionally drowned in the noise from the streets outside when the door was opened. I wondered, was she gazing rapt in adoration at the Host in front of her. For all I knew she could have been meditating on the huge blurred happening outside. I waited until another nun, also enveloped in white, appeared through a side curtain and, after joining the first one in a brief ritual, replaced her, silently.

Next day was Sunday, the first after Easter. I went into the cathedral and joined a small group of people behind a boundary cord supervised by an usher as a mass was proceeding far above us in the cavernous interior. A hymn was being sung. When it was over, a madwoman suddenly appeared down from the congregation, clapping and waving her hands into the air. The usher was affectionately firm with her as he shooed her out, parrying her attempts to embrace him. I made my exit after her and boarded a bus outside, back to the Alhambra. I headed straight for the church that was built where the mosque had stood for centuries. A mass was just about to begin. There was just a scattering of others at the top of the church. I sat at the back. The big doors were closed behind us. The priest appeared in his vestments and set about the familiar ritual. I let the unknown Spanish words wash over me. Soon he had embarked on his sermon and was in no hurry. It seemed as if he repeated a number of times the name Thomas; I guessed that doubt was his subject, and his demeanour indicated he wasn't in favour. He didn't seem disheartened by the impassive faces of his tiny congregation; he entered fully into his own flow. His voice rising, the staccato of his words carried him along against the background of the echo of his previous ones still resounding in the almost empty church. Eventually, it seemed as if his energy was ebbing, but his voice rose for one final flourish before he gave way and carried on with the set ritual: the theatre of the consecration. When he reached the giving out of holy communion I retreated to the outside world.

I went to a bar/café across the road. The barman was friendly, and the small clientele had an easy camaraderie among them; it seemed as if they were all on brief breaks from their involvement in one way or another in the Alhambra tourist trade. It was like being back-stage. One man in particular was talking jovially to a younger couple, and there was banter in Spanish between them. It seemed as if he was making fun of the foreign tour leaders that he had to deal with. As the couple were leaving he followed them to the door and as he shook their hands he repeated a few times in English with a sardonic tone, "So nice to meet you". He disappeared after them. A while later I saw him again, outside the Alhambra. He was in the midst of a group of tourists, talking to their leader with a graciousness to the point of obsequiousness. I wondered, if he were called up before a tribunal set up by his peers? By his employers? By God? To which identity would he swear allegiance?

I left the Alhambra behind me and got a bus up the opposite hill of the Albaicin. When I reached the plaza of Saint Nicholas it was already crowded with hawkers and tourists and a few musicians playing flamenco. In the middle of it all sat a gypsy woman selling her castanets, clapping one of them impassively to the music. Her face lacked all enthusiasm, and had the look of someone who has seen too many things down through the years; she didn't see me. I went in search of a mosque that I had seen being built a decade before, the first to be built in Granada for centuries. A short search in the vicinity yielded it. The mosque itself was closed to visitors, but the garden was open. In the garden I asked a man who seemed to be caretaker was he aware of a Muslim cemetery in Granada. His English wasn't adequate to understanding me. He went away, and came back with a leaflet in English about Muslim sites in Granada, but there was no mention of a cemetery. As I left, I heard the call from the mosque's minaret, faintly.

I came down a different way to the river right underneath the Alhambra hill. I walked along it. There were many young people out in the evening air, many with rings on their faces, some with blankets in front of them arrayed with trinkets, some playing guitars, some romancing; some drinking from bottles of beer, and occasionally a familiar waft of something heavy and sweet. I came across one young man at one of the old small bridges, with a sign saying he was giving poems away for free. There were pieces of folded paper in front of him and a receptacle for donations. When I asked for one he typed a few lines on a typewriter and he folded the paper and handed it to me. I thought how, if one were young and with a full heart, Granada is the place to be. I with my old heart carried on.

Back in the square, most of the traders and hawkers had finished for the day. I saw the balloon-seller I had seen earlier, dressed as Mickey Mouse. This time the head of the costume was lifted up, revealing the small serious face of a South American man, talking intently to another man of the same nationality. The contrast between the expression on his face and on the mask was startling. Before going to my room, I stood outside on the balcony of the hotel. I noted the ramparts of the Alhambra. I thought how, in the heyday of the caliphate, the commanding prospect from there must have had its corollary commanding presence to the commoners living below.

That night for much of the time sleep eluded me as I wondered, perplexed: after that trio of assays on a boundary, where now can this journey and this writing go? The next morning, on the internet in the lobby of the hotel, gradually I came round to searching for the Muslim cemetery in Granada, and eventually I found a reference to it being part of the municipal cemetery, the whereabouts of which was given as a short distance past the Alhambra. I took the bus back up, and carried on along the road past it. Very quickly I had reached the municipal cemetery. A notice beside the gate, mentioned different sections, called patios; off to one side, separate from the others, was mentioned patio islamico. Another notice on the gate spoke about the cemetery providing a journey through time. I went inside.

It was an ordinary day in the life of a graveyard: lingering near the gate was a group of people in the aftermath of a funeral, some of them wearing black. One woman was crying; others were comforting her. Inside were workmen about their various roles, tending it. I walked the straight avenues with buildings either side filled with those interred, rising on each side round me, rather than in Ireland where the graves are characteristically below. A couple of men passed by with ladders, presumably having needed them to climb to leave a tribute at the tomb of their loved one. I walked until I came to a dead end. I returned to the gate and mustered the wherewithal to ask a woman behind a counter, first apologising with the word, English, pointing to my chest, and then asking her, patio islamico? I didn't understand her words, but did her gesture as she pointed outside and back around past a building that was the crematorium to a roadway behind. The road led above the crematorium, from the chimney of which a faint plume of smoke was rising. As I walked along, the road quickly became a dirt road that a sign announced to be a camino. Beside the road I saw an old condom in the grass. I remembered Marcus Aurelius on the transience of all mortal life: "yesterday a drop of semen, tomorrow a handful of spice and ashes." Soon I came to a walled area with a gate. The gate was closed. A sign said: Cementario Larawda de Granada. There was a small track around by the wall. I walked it. The graveyard looked unkempt compared to the other one. Most graves that I could see looked dishevelled and quite old, many of them unmarked. I walked around to the very back along by a high fence, wondering might there be a way in. I stopped at the back looking out in front of me at where the hill fell in a precipitous descent. There was nothing but scrub on the side of the hill. At the bottom were some houses, then a busy main road. Lines of hills receded beyond that, until they became mountains, reaching back to the upper reaches of the Sierra Nevada. I looked around through the fence into the cemetery and to my astonishment it was the first thing I saw: the ultimate grave; the gravestone had the name on it, written in English: Muhammad Asad, born 2 July 1900, died 20 February 1992. I lingered looking open-eyed at it, amazed that I had found it.

Reader, I do not anticipate that you will know the name Muhammad Asad. I had never heard the name until a couple of years ago when I chanced upon a documentary on Al Jazeera about his life, made by an Austrian, George Misch. There is no need for me to recount his biography; a brief perusal on the internet will yield far more than I could give. A few details are sufficient for my purposes here. He was born by the name Leopold Weiss. He was brought up in the same milieu as Franz Kafka: a Western Jew in the dying world of Mitteleuropa at the birth of the twentieth century. Unlike Kafka, he was part of the Jewish tide into Jerusalem. Around the age of twenty he went to live there with his uncle who had a house just inside the Jaffa Gate. It was there he had an illumination that changed the direction of his life. He suddenly realised that the Arabs he saw around him in the old city were far closer to the world of Abraham than were the Zionists who had come from Europe. Thus began for him a journey into Islam. In the middle of his life he wrote an account of that journey. The book was called, *The Road to Mecca*. For a while he was at the political and religious centre of the Muslim world. Later, there was a disaffection. He lived for a couple of decades in Tangier, engaged in what he had come to see as his life's work of translating the Koran into English. At last he crossed the short stretch of water from there back to the threshold of Europe: Andalucia. He lived near Malaga until his death in his nineties. Asad was in his fifties when he wrote his autobiography. In it he repeatedly uses the metaphor of the crossing of a bridge over a chasm to characterise the journey he had made from the world of his inheritance to the world of Islam; a bridge so long that he had had to reach a point of no return before the other side became visible. The final sentence of the book imagines him riding in the midst of thousands of his new brethren. Looking behind him he sees "the bridge over which I have come: its end is just behind me while its beginning is already lost in the mists of distance." The film documentary on his life, in an echo of the title of his autobiography, is called, *A Road to Mecca*. The final scene of the documentary shows an altercation in the cemetery between a Muslim cleric and a man who introduces himself as Asad's assistant. The cleric is complaining about the lack of orthodoxy of Asad's gravestone, a stone's throw from where I was standing.

Later that day I wandered around the alleyways of the city. I found myself once again in the square beside the Iglesia de San Gregorio Betico – the church of the white sisters. When I saw the door open I went into it. It was shortly after six o'clock and there were about ten of them assembled, enveloped in their white habits. Their voices were all that was revealed of them. I waited as they rhythmically said their prayers. Then they began singing, fervently. Those white sisters: why should their voices not be pure? I decided to wait until they had finished their singing and praying. It took about an hour before a couple of them, who by their set and movement looked older, made their exeunt through the side curtain. The others followed them, leaving one sister in still silent solitary prayer.

Back in the hotel, before going to my room, I stood on the balcony looking down at the square. Its day's crowds had receded. Traders were shutting up shop. In the middle of the square I saw the Mickey Mouse man, standing still and alone. A child came up close to him and stood looking at him, but it seemed he paid no heed to the child. The child was called away then, and the Mickey Mouse man was a lonely-looking figure as he stood awhile longer, as if he was silently considering. Eventually, he slowly made his way to one of the alleyways off the square, and disappeared.

October of that same year saw me back once again in Granada, in the lobby of the same hotel, Los Tilos, in the Plaza Bib Rambla. Soon after checking in I was heading back up the road past the Alhambra. I walked on, across the roundabout to the main cemetery. There was a crowd gathering around the crematorium. Most of them were young people, and I imagined the deceased person was also young. I walked around along the road that soon became the dirt one that was announced as a camino. Looking down I could see the crowd, and a light-coloured hearse, and the crematorium waiting. The gates of the Muslim cemetery were again closed. I walked the way I had gone before around it, and soon I was standing again looking in through the fence at the grave of Muhammad Asad, still, exactly as the last time. The sun was hot. I sought shelter beneath a tree and looked down over the valley to the mountains beyond. Meanwhile his grave was quiet behind me. The gate was still closed when I walked back round. I carried on down past the crematorium below me. The cremation ceremony was over; the mourners were gathered outside again in a kind of alcove. I saw none of them look up to where I saw clearly the smoke rising from the chimney. If one hadn't seen the surroundings one could be forgiven for assuming that it was an ordinary social gathering. I saw some embracing long, tightly, and affectionately. Meanwhile, the smoke had all but died away, and most of the gathering had left or were leaving. I loitered around the Alhambra until evening. It was almost dark as I made my way among others into the Generalife Gardens. There was some light still left in the sky above the palaces. The gardens were lit only with small lamps along the pathways. The twilight air was resonant with crickets calling. It was only as I was leaving that the almost full moon began to appear through the tops of the trees. Back at the hotel, from the balcony, I watched as the moon rose, floating bright, making its silent voyage through the darkness. I retired to a night divided between intervals of waking and of sleep filled with a forest of forgotten dreams.

Next morning I returned to the roundabout beyond the Alhamba, which is the boundary between the tourist world and the other world. As I approached the Muslim cemetery I saw a motorbike parked outside, and a man who disappeared inside. The gate was open. When I asked the man could I go in he directed me to a younger man. When I mentioned Muhammad Asad, immediately the older man grinned broadly. The younger man was more reserved, with a Muslim air about him, for which I imagine I share with Asad an instinctive empathy. He asked me did I know where the grave was; I said I had seen it from the outside. I went down to it. The gravestone was supported erect by concrete spread round its base. I stood beside it, paying my respects. I placed my palm on the warm stone. I picked and pocketed the lowest cone from a cypress tree that overhung it. Again I noted the dishevelled state of most of the graves as I walked back through them. The younger man was waiting. I thanked him, and asked him did many come to visit the grave. He said people come from different countries, and then, as he made as if to begin listing them he hesitated, into which I volunteered my own country, Ireland. He told me the cemetery was from the time of the Spanish Civil War, that about fifty Moroccans had been buried there; he mentioned Franco. As he told me the story of it I couldn't make out had they fought for Franco or against him, and I didn't ask for clarification. What I gathered was the most important: that the men buried there were killed in other men's wars, fighting for their next meal, like so many men before and after them.

The young man was locking up as I was leaving. I headed out the camino. A number of cars passed me. Most of the front seat passengers seemed to be consulting their personal, electronic, communication devices that are the successors of mobile phones. The cars were throwing up dust in their wake. I turned off the road away from them, along a narrow track, over a broken bridge, into a quiet solitary mountainscape on which I met almost nobody. A camino with no particular destination. I thought of Asad's road: how all the other roads had fallen away leaving the road that had come to this: his road that brought him back to Europe, and came to an end a short stretch of road past the Alhambra. I remembered the challenging clarity of the question in my own adolescence: how does one live one's life? A question whose answer we can never know beforehand as each of us takes our whole life to answer it. I thought, amidst the cacophony, to listen out for a tone, and to follow it, would be the height one could hope for.

That evening I returned through the crowded alleyways to the church of the white nuns. Outside, in front of its porch, there was a band playing music that was not religious, and what they were playing they were playing LOUD. Inside, the nuns, about eight or nine of them, were singing their vespers, with one of them at the back playing the harmonium. They seemed to be ignoring the noise coming in from outside. What could they do but carry on? They were repeating what I had seen and heard in April, until again, just after the clock rang seven chimes, the older ones made their exeunt, followed by the others, leaving the one who had been playing the harmonium kneeling in silent solitary prayer. Outside, the band had finished, but the square was raucous. A band of young people came along in a pagan parody of a religious procession, singing and dancing in revelry. I wandered in the midst of the revelry before returning to the church again. The sister was still silent and still, with the ticking of the clock, and the sounds from the outside let in with the opening door. With what ears did she hear it, the noise of the world? Had she too, like Asad, abandoned the name she was born with, her life given direction by the charge of belief, the ritual of her days given meaning by a book? I was standing at the rails when another sister appeared and replaced her. Before she left she quickly came back to retrieve her hymn book and, in the enchantment of the moment, I imagined it was a young Arab woman coming towards me.

Back outside, the loud streets of Granada were in an unequivocally festive mood. I made my way back through them to the Plaza Bib Rambla. It was crowded with families: some were queuing at a tent that functioned as a planetarium; there was a mechanical roundabout for children powered by a man pedalling a bicycle; there was a couple blowing huge bubbles to the excitement of young children, and in the middle of it all I saw the Mickey Mouse man. This time he had company, a woman of the same nationality, even smaller than he was. It seemed that he had adopted the role of a mascot, to be photographed with children, as she was entrusted with the selling of the balloons. She was holding many of them, of many different representations, of cartoon characters and animals. She was having difficulty managing them. As she tried to fix them, a horse at the end of a string became dislodged from her bundle and disappeared up beyond the buildings into the dark yonder. Among the revellers, I saw no one look up. Then arrived in the square a troupe of flamenco dancers and musicians. The dancers soon began their dance. Their fixed smiles were a counterpoint to their flowing steps as they took their lead from the music. The people crowded round them. The Mickey Mouse man and his companion were left out on the sidelines. I saw him lift up his Mickey Mouse head and discuss with the woman what they should do. What could they but carry on? After a few tunes and dances the dancers and musicians left the square. I too soon left while the square was still in full flight, with the Mickey Mouse man and his partner still in full flight among them. In the hotel I stood out on the balcony above the noise and crowds of the plaza, and watched the full moon rising silently into the sky above the Alhambra and the city of Granada, shining silently as it did centuries ago on the old caliphate, as it will presumably continue to shine, equally and without prejudice, from orient to occident until the end of the earth.

The next day out in the square I was standing watching the man fold up his mechanical roundabout onto a trailer when I caught sight of the woman balloon-seller again. She was walking quickly through the square with her balloons, and then she disappeared down the same alleyway where I had seen her partner disappear in April. I followed her and saw her going in a doorway. The door was open and I followed her. She had disappeared up a stairs. I stood in the twilight of the bare empty hallway listening to the scraping of the balloons being squeezed through the narrow space of the stairwell above me, until a door closed. That sound seemed to signal a silence. It lasted no more than a moment before I began to hear again the street-sounds coming in through the open door. The noise of the world: I went back out to it.

THE SILENCE OF SEAMUS HEANEY'S FATHER

Seamus Heaney died in 2013 as Summer was giving way to Autumn. His funeral Mass in Dublin was broadcast live on television. It was apt, measured and moving. His fellow poet Paul Muldoon gave a eulogy, in which he used a motif of the letter B. He brought it to a close by speaking of Heaney's bardic beauty and, finally, the beauty of his being. Although no judgement is clear and simple on any human life, the image portrayed at his funeral mass of the being of the man, Seamus Heaney, was admirable and appealing; it was one of someone at harmony with himself and with those around him. God rest him.

The programme finished with the end of the mass and, as the mourners left the church, the cortege was preparing for the final journey to his home place, Bellaghy. As the cortege was heading north, I was driving out west, to Mayo, to cross to one of the small islands in inner Clew Bay. My departure had been precipitate, and the journey had an element of flight about it. It was only as I passed through Westport, when the sunshine that had prevailed for most of the day gave way to drizzle, that I realised that I had left my normal glasses behind me and had brought only the prescription sunglasses I was wearing. As I drove down to where my currach was moored, Heaney's play, *The Burial at Thebes*, was coming onto the car radio. I parked and walked in the drizzle to the mooring where the currach was waiting. It was misting heavily on the short journey across. I could barely make out the island mooring. I landed, pulled the boat out and headed up to the old schoolhouse. I lit a fire, and listened timorously to the rest of *The Burial at Thebes*. I stood outside for a little while when it was over but, between the drizzle and the dark glasses, I could see only obscurely the regular flashing of the lighthouse on Inishgort, and inside it the erratic flashing of the light at the jetty that used to light the way for Sean Jeffers, that now lights the way for no one. In the schoolhouse I lit candles and focussed my attention inside for the rest of the evening. Before bed, I read Heaney's late poem, "Uncoupled", that I had downloaded onto my laptop. It has two deeply fond vignettes from the depths of his childhood: of his mother in the first verse, and in the second of his father. The poem ends with the memory of his father, in a loud confusion of cattle, turning his attention away from his eldest son to the adult world around him: the first speechless loss of the poet, that the poem six decades later redeems.

I woke a number of times during the longish night with the uneasiness of an animal alone in the dark. I dreamed the house of my own childhood was still standing and that years after his death my father had returned and was living his life as an old man. We were all celebrating the miracle of it. I caught his glance just before I was leaving and basked in his warm approval. Then, as I went out the door I cast another glance back for a further glance from him, but his attention was deflected by the crowd around him. I woke, holding on to the memory of the image of the previous glance. Although I couldn't remember the occurrence, it seemed that the dream was the echo of a memory.

The figure of the father is fundamental to the poetic dimension that Seamus Heaney retained of his religion. He alludes to this in the series of interviews he gave to Denis O'Driscoll, in the closing chapter of the book, called *Stepping Stones*. He was responding to two quotes from Wallace Stevens about the close kinship between religion and poetry. Stevens had said that God is a symbol for something that could as well take the form of high poetry. Included in Heaney's response were the following three statements: "poetry is the ratification of the impulse towards transcendence"; "poetry represents the need for an ultimate court of appeal"; and, pace Pascal's dread of the eternal silence of the infinite spaces, "there has to be something more than neuter absence." Having lost the transcendent faith he had inherited from his father, he saw the making of poetry as the ratification of the impulse. His mention of an ultimate court is a reference to his poem, *The Stone Verdict*; he goes on to quote it: how his father would expect more than words in the ultimate court he relied on through a lifetime's speechlessness. He said that he imagined the celestial silence beyond us as a divine corrective to human protestation: the silence behind the silence of his father, which his own lifetime's profession had breached.

The morning after Seamus Heaney's funeral I went walking on the island. The sky was lightening. There were pools of pale September sunlight scattered around the bay. From the hill of the near headland I noticed two in particular, one around the lighthouse on Clare Island, and the other around the near lighthouse on Inishgort. In front of the lighthouse I could see clearly the house of Sean Jeffers, the last of the line of men who in their time had tended it. To the south, above the bay, Croagh Patrick was shrouded. Nearby, on the tarbert, I saw Brendan Joyce cutting rack: a familiar figure in the bay for decades, he had disappeared for some years during the mad dream of the first decade of the millennium, working on the mainland during the economic bubble, driving a lorry for a builder. Now he had returned, and his distant figure, bent down, silently cutting the rack from the rocks, was once again a deeply characteristic image of the life of the bay. Later, as I stood in the schoolhouse doorway, I saw him passing in his boat, pulling his bale of rack, waiting for the tide to rise before he brought the harvest of his day's labour into the shore at Rosmindle. Our salute was the prolonged and emphatic one between boat and shore of good will at a distance.

Autumn of 2013 gave way to Winter. It was a mild one, most memorable for the storms, and for the sea surge in January 2014, which caused devastation along the west coast of Ireland. Cars were washed off the pier at Cleggan, and a bridge near Rosmoney was swept away by the flood tide receding. A month later, at the following big strand, a few of us walked across the tarbert to the island. The big tide had divided the island in two, and had left seaweed for the first time inside the gate of the schoolhouse. We had an hour there on either side of the low tide turning. On our way back, I saw Brendan Joyce off at a distance, cutting rack. I had a hailed exchange with him. We bemoaned the storms. He shouted that the sea had risen five foot above the road down to our common mooring, and he said the likelihood was that there is worse to come. I shouted back that it was as well in that event that we have boats. The wind got up then and I couldn't hear his reply, if there was one, and immediately I regretted my facetiousness about something of much more immediate material concern to him than to me.

A while later I joined a gathering of old cronies in a pub. All of us shared in common that we had been born in Dublin in the middle of the last century, and had spent the bulk of our adult lives in the West of Ireland. There was talk of the winter floods, and of the question of global warming. There was a consensus that we were unlikely to see a global catastrophe in our lifetimes. We remembered our shared past: how our lives had been greatly privileged in their stability and economic security, despite the turbulence of the wider world. We acknowledged the transient bubble blown up by the suburban Dublin of our adolescent world as the nineteen sixties gave way to the seventies, with the sudden impact of the wider world. There was mention of the death of Seamus Heaney: in one of his interviews he had spoken of being fully formed by 1963; there had been for some of us a childish tendency to see him as belonging to a previous generation. Although in awe of the fecundity of his wordsmithery, he had seemed to be part of the Irish Catholic world we were turning our backs on. We remembered our assault on the axioms of our inheritance, and how we wholeheartedly applauded when the man sang to the mothers and fathers throughout the land, not to criticise what they couldn't understand. We remembered how we saw ourselves as on the cusp of a new world, until our worlds became amorphous.

Winter gave way to Spring. A good spell of weather dried out the land, and in the nature of things the real presence of the winter floods sank back in people's memory. Out of the blue I found myself back on the island on a balmy May evening. I sat outside the schoolhouse watching two herons on the mouldering gables of the ruined cottage close by; they looked to my eyes like oriental monkeys. I walked over the shore to gain a closer look, but they languorously lifted themselves and flew off at my approach. As I stood by the gable I remembered Seamus Heaney's poem, "The Stone Verdict": after the first verse, with its Christian image of the final court, the second turns to a metaphor from the ancient Greek world and imagines the father's apotheosis in the silence of such a ruined homestead. The silence of stone. My evening in the schoolhouse was wordless. I paid a brief visit over to the beach at twilight. The water was still, and there was silence but for the rhythm of a slight wash on some distant shore. The lighthouse had begun flashing, and the light above Sean

Jeffers' jetty was still going on and off without rhyme or reason. I went outside in the middle of the night and saw the dark vault studded with crystal. In the middle of it all, directly above the schoolhouse, there was the amorphous glistening of the centre of the galaxy that one small island of life has named the Milky Way.

At dawn the cloud cleared into a beautiful May morning. There was a cool breeze but the sun was shining. Swallows and starlings were flitting around the house, and on the hill the larks: the larks were rising. I saw Brendan Joyce coming out of the mouth of Rosmindle. His figure and his boat were a glimmering shadow in a huge flood of sunlight. I marvelled his dedication and commitment: his long day of hard silent physical work on a Sunday as I sauntered the near headland. The view was expansive, deeply familiar. One lark rose straight in front of me at the top of the hill, singing its heart out as it climbed almost perpendicularly, until it drifted a little before it glided downward again, landing where I could see it again right in front of me. Half-way down the far side of the hill, as I was walking past a thicket of brambles and gorse, I heard a bark, and when I looked I saw the light brown reddish hind quarters of a fox disappear into the undergrowth, presumably its bark a warning to its own kind on seeing an intruder.

The bark of the island fox: it seemed a summons to a laudate; a sound beyond the human world, taking me beyond the thicket of thought. I sat on a bank looking out. There was thin cloud overhead, but Clare Island was bathed in sunlight. Above the bay the Reek was dark but the church on top of it was gleaming. For moments I gained glimpses of the purity of the manifestation of the world of the islands around me. I remembered Seamus Heaney's comment about how there has to be more than neuter absence in the vastness; I thought how the flamboyant light and life and colour that I saw around me was as close to the centre of the universe as anywhere else – as intimate – as intrinsic.

As I was heading down to the shore, Brendan was coming along slowly in his boat, pulling another bale of wrack behind him in the rising tide. He was taking it slow, minding his load. I followed him in the currach. He asked me had there been a lot of water in the currach as there had been a woeful amount of rain the past week. I asked him had he cut all of the bale of rack that day and when he said he had I complimented him on the hard work he had done on a Sunday. He asked, "do you think, the Man Above, do you think He'll let me in?" I said I thought He would. He laughed, "do you think He'll take pity on me?" I assured him I was sure He would, and our leave-taking was a shared laughter as I pulled on past him, in through the mouth of Rosmindle, back out to the wider world.

MITILINI HARBOUR

When the plane landed in Athens airport, I was soon outside with the inevitable confusion. I was helped by a few others to find the way to the platform of the metro that took me in. As it came in near the city centre, it crowded with Athenians. Many got off at Syntagma Square. I waited for the next stop, Monastiraki. Outside on the street I asked a man with the one word, "Athinas?" He pointed to the street, and I walked the short distance to the Attalos Hotel, gladly relieved to have forded the first part of the journey. The welcome from the man at the desk was cordial. He gave me the card to my small good room. I visited it briefly before heading up to the roof-top bar where I drank a beer and sat looking out at a building on a hill, wondering was it the Acropolis, until I turned around and saw the Acropolis, indubitably. The Parthenon was lit on top of it. I sat on, drinking in the powerful sight of it, and

I was still sitting there at midnight when the manager announced that the bar was closing.

I had been in Athens a few times before. The first time had been in 1980, at a time of personal crisis, under a siege to my fragile sense of identity, overshadowed by the question: how to go on from there; how to live? Decades later, in 2016, having reached a stage when I wondered might I not ever see Athens again, I was glad to be back. I returned to my room and, after all the nerves, slept peacefully and sometimes deeply.

The next morning I headed off walking with the image of the Acropolis periodically in front of me, memories stirring. Then, when I was underneath it, the place came back to me when I saw it. But the ticket office was shut. There was a notice saying that the Acropolis was closed as part of a strike of public services. I loitered around the entrance as various groups of people came along and were disappointed. I purchased in a stall a frozen lemonade and nursed it as it gradually melted. I sent texts home to fond figures of the world that had grown up around me and within me since I first stood there. I was about to leave to go down when I heard a man mention that the new Acropolis museum was nevertheless open. The man in the stall confirmed it. I walked on down to it: a new airy building, more spacious than required for the amount of artefacts currently inhabiting it. The atmosphere was friendly and congenial, and the assistants were helpful. One of them pointed me towards an exhibit I had seen in the guidebook: a votive offering in a cranny on a pillar from the sanctuary of Asclepius: a pair of eyes. I had a snack outside in a balcony café. Sitting there with the incontrovertible reality of the Acropolis rearing above me, it was easy to see why the artefacts dubiously appropriated by the British during their colonial heyday should be returned by them forthwith.

I wandered along a road that I had a vague memory of leading to a hotel I stayed in years before, but I didn't find it. I passed a man who was playing a stringed instrument. I wondered was the music Syrian as I gave him a few coins, and I sat secluded in trees listening to the music, drawn once again to the lure of the Near East. The streets emerged into familiarity as I made my way to Syntagma Square. I crossed it to the parliament building. It was with a feeling of inevitability that I went in to the national gardens, remembering 1980. As I sat there I remembered the orange trees, but had no intimation of the desperation that I had felt way back then. That time had become so remote as to be hardly real any more. I imagined what my thirty year old self might have thought if he had passed that way, his youthful eyes seeing the figure of his fate as an elderly man dozing on a park bench. Beyond the imaginings, the incontrovertible: how one has actually lived.

My thoughts moved to contemporary concerns. I returned to Syntagma Square where a friendly bus driver directed me to the bus stop for the airport. There was a notice giving the times. I was about to take out a pen to make a note when a couple came along, and the man took a photograph of the timetable, giving me the example that I followed. I then made my way circuitously back to Monastiraki metro station where I saw that the earliest metro to the airport would be suitable to my purposes the next morning. Back in the roof-top bar as the day darkened the lights came on again on the Acropolis. I heard the chanting of a demonstration, and could see a crowd on a street nearby. I went down immediately, but caught only the periphery of what seemed a modest demonstration that the city took in its stride. I thought, despite the crises, democracy is unquenchable in its birthplace.

The next morning I was up and out early. I walked along as I had the morning before, and soon I was at the open entrance of the Acropolis. Familiarity once again resurfaced as I climbed the steps leading up to the Parthenon. It still looked like a building site, which it has been, it seems, for most of modern times; but the spirit of the place came through again, until the crowds arrived and took over the foreground. I sat a long while, inarticulately imbibing the atmosphere of it, gaining no clearer understanding of the world of those for whom it was perhaps the most important part: the sanctuary of the gods, while, down below, Socrates was loitering in the market-place, acting as midwife at the birth of secular rationality, until he was indicted for impiety.

I headed back down to the world below, into the ancient agora. I made first for the museum, and the stoa of Attalos. I sat outside underneath the stoa, close to a bust with the grave preoccupied face of the emperor, Antoninus Pius, Marcus Aurelius' adoptive father. I went walking, lingering. It came as a surprise to recognise the large area of scrub ground and trees. I moved very slowly. At one stage I was standing and saw a slow movement on the ground. It was a tortoise. It stopped, looking round at me, before it carried on its patient journey. I climbed to the temple of Hephaistos, and sat on a bench close by it. It was shaded from the sun by a tree, and afforded a good unimpeded view of the Acropolis. I sat for what seemed a couple of hours there in reverie, wondering wordlessly about the ethos of the Acropolis. I fell asleep a few times, waking once to the words: "a beautiful lost thought." Those words were all that remained.

That night was a night of short fitful sleep, with numerous wakings, before the call from reception at five o'clock. Soon, I was down, handing the man my card, and I headed out on the street. It was still alive, with a mixture of people who were still lingering in their night, and others just beginning their day. There were a few stall-holders who seemed as if their day never ends. I went down into the metro station and began my journey to the airport. On the way I had a series of minor mishaps of no lasting consequence; suffice to say that I was bothered and bewildered by the time a woman behind a desk casually and calmly handed me my boarding card. I made my way to the gate with my fellow passengers, two of whom were men flamboyantly dressed in Arab clothes. A bus took us out onto the tarmac, and stopped at a propellered plane. It rose mistily into the sky. I could make out little of the Mediterranean until it began to descend, and I caught my first sight of the island of Lesbos.

I had never been on Lesbos before. The images in my head were all from the news reports over the previous year of boats landing, filled with refugees. They came in their hundreds of thousands; thousands were drowned. It was called Europe's refugee crisis. The islanders welcomed them and helped them, assisted by volunteers. There was a while when Europe rose to the challenge, particularly Germany that took a million refugees during the course of 2015. There was a lull in the arrivals during the winter. But soon after, by the beginning of 2016, the flow was beginning to build again, and again there were dramatic scenes on the news. I was particularly struck by a video I saw on the "no comment" series on Euronews, one evening in January, after I had booked my flight to Athens; there was a boat filled with Afghan refugees being rescued by a team from Medecins Sans Frontiers and Greenpeace. Amidst the *rira and ruaille buaille*, a woman was soothing her crying baby by gently tapping its mouth in a maternal gesture of great tenderness. As winter gave way to spring, it was beginning to appear as if the flow of refugees in 2016 would surpass that of the previous year. Fences began to be erected at the borders of countries north of Greece. The gates of Europe were closing. The main refugee camp, at Moria, on Lesbos was turned into a detention centre. By the time I arrived there in early April, the European Union had made a deal with Turkey, using it as a buffer, without putting under too close a scrutiny the tactics that Turkey was using. NATO ships began patrolling the Eastern Aegean. The passage of boats across the narrow straits was almost quelled. One flank of Europe's refugee crisis had been assuaged. The crisis of the refugees continued elsewhere.

The plane landed at Mitilini Airport, right beside the coast. Soon, most of my fellow passengers had departed in hired cars or taxis. I loitered for a while before crossing the road to a bus shelter where a woman and a man were already separately waiting. The wait was a long one, but I didn't mind, as I stood there taking in my new situation. The airport was quiet in front of us. The sea was directly behind us. On its sparse rocky shoreline, I could see what looked like the remnants of two dinghies, along with some flotsam and jetsam scattered along the shore. Eventually, a bus appeared from the direction of Mitilini, turned and came back to us. The driver had a modest friendly air. The atmosphere was parochial. As we approached Mitilini, he drove off up a hill through suburbs and, in a circuitous route that included reversing turns, he picked up his quota of passengers, many of whom knew one another. We returned to the coast road, and soon Mitilini harbour appeared in front of us. In the middle of the waterfront I could make out the hotel I had booked: the Hotel Lesvion. We passed it and the bus stopped shortly after. A woman came up behind me who told me that was the last stop. She introduced herself as Jenny, a Catholic from Wales, and she invited me to come with her for tea to her church which was just around the corner. I went with the flow. After all the nervous imaginings about my arrival in Lesbos, I could never have imagined the utter eccentricity of the individual experience that presented itself. She opened up the small church and brought me to a side room where she served me tea and savouries that she had bought, and she talked lengthily, meanwhile apologising for doing so. We spoke of course about the refugees, and of the Pope's visit to the island the following week. She told me of a mass that was to be said by a bishop in the church the following Monday. I said that I would probably attend.

After checking into the hotel, I went wandering, along the harbour, past the docks for the ferry boats, and out beneath the battlements of the old castle of Mitilini on the hill above me. There was a refugee encampment behind a beach underneath it. There was a van marked canteen, and a sign saying "No Border Kitchen". I exchanged greetings with a few young Asian men on a wall, one of whom said "kalimera", then corrected himself with a laugh, saying "kalispera". I tapped my chest apologetically and ridiculously, saying "English, English". I was standing looking out into the harbour when there was a slight movement behind me: it was a young boy of perhaps seven or eight; he held out his hand begging in a way that showed that he hadn't been reared to it. I dismissed him and immediately he retreated. But then I looked back and saw him joining a young pregnant woman who was obviously his mother, and an infant in a pushchair who was presumably his brother. I went back to them and handed the boy my miniscule donation.

I had been told that during the height of the refugee influx there had been people sleeping all around the harbour, including right in front of my hotel. But by the time of my visit the populace of Mitilini seemed to be taking it all in their stride. The streets were crowded, and the atmosphere was voluble and noisy. The cafes and bars were crowded with Saturday evening proceedings. I went into a place that had an element of fast-food about it. Outside I saw some young girls running, exuberantly playing. One of them came in and asked for something from my plate, until the proprietors shooed her away.

There was a scattering of other people down at breakfast, including a few Greek soldiers. Afterwards, as I sat out on the balcony with my coffee, one of the soldiers came out for a cigarette, and greeted me with a hesitant good morning. On the street below, a band of soldiers marched past to the slow beat of a drum. They had disappeared by the time I went out after them. There was a Hellenic coast-guard boat moored in front of the hotel. Close by it was a rescue boat that had the poignant human presence of life-jackets scattered around its deck. I walked as far as an Orthodox church. It was closed, but the porch was open, and a woman was there before me, lighting a candle which she then stuck into a bed of sand that was there for the purpose. I did likewise.

I headed out to the marina. There was a clubhouse with a sign that said Marina Yacht Club. From my vantage point in the doorway, the atmosphere, the company and the décor were indistinguishable from a similar gathering in a yacht club in Ireland. I left behind their hubbub and carried on out the pier. The waters were choppy enough in the harbour, and choppier beyond it. On the pier there was a group of more than a dozen people beside a boat marked, "International Maritime Rescue". They were dressed accordingly. A mini-bus came along and took most of them away. There were just two men left, and I plucked up the modest courage to approach them. In voices that sounded German they confirmed that the crossings had hugely diminished following the deal between the EU and Turkey. Their communication was formal and reticent. I commented on the weather, saying it would have been rough for a crossing, and the older man spoke of a swell warning that had them on dry land for the day. I retreated past the yacht club, back to a small pier at the edge of the inner harbour. It was crowded with groups eating and drinking. I chose a bar and ordered beer. I sat outside directly beside the water where a couple of small fishing skiffs were bobbing up and down in front of me. A young refugee boy came along selling tissues. At first he immediately retreated to my refusal, and then when I bought I could see the relief on his face that showed his shyness matched mine. There was a woman and man at a table beside me, who seemed to be deeply unhappy, who, after words, had become silent. Two men were in voluble conversation at a table behind me. Musicians for a while struck up at a nearby crowded table. A pale delicate child came down briefly and looked into the water.

As I went out past the ferry terminal, there was a huge ferry tied up there that had a notice that it was to head off that night to Athens. There were two young Asian men in the shade of a vehicle parked on the quay, scrutinising the deserted gangplank with an intensity that they seemed to try to hide as I passed. At the end of the pier I sat awhile beneath a covered pavilion. Across at the other side, where I had been earlier, I could see Sunday sailors in their small yachts sailing seemingly thoughtlessly around the mouth of the harbour. On my way back I could see the No Border Kitchen encampment on the beach, from where the inhabitants must have watched the comings and goings of the ferries. Along the rocks beneath the pier there were numerous pieces of detritus of their own much more fragile voyages: the remnants of a dinghy, lifejackets and rings, and nautical bits and pieces. I passed a man and woman who were employed cleaning the environs of the pier. They had a sweeping brush, and a big bin on wheels. Amidst the detritus in the bin, all my eyes were able to fix on was the sight of a single shoe.

Later, I went wandering again, back out the pier in the darkness. There was a handful of stars in the dark sky, and a crescent moon. The wind was still up. The water was like a living thing. The dark water: I could only make out its dark undulations as I stood there looking across at the glimmering beyond it of the lights on the coast of Turkey. Somewhere out there in the dark water was the boundary between Europe and Asia that will always remain invisible, by night and by day.

I came back past the ferry terminal, and the municipal tourist office that I never saw open. There was a British naval boat moored at the quay, with its notice: Border Force: those two words that historically have gone too often together. I thought of the contrast between the present migration of refugees and the migration of the European colonial powers over the past few centuries: how the European entry into the rest of the world characteristically began with invasion, was followed by the subduing of the inhabitants by force of arms, and afterwards the drawing of borders. I thought of the images of the refugees making landfall on Lesbos, defenceless and vulnerable. I remembered the video I had seen in January: the image of the woman and her child that transcended every national boundary.

The next morning, I walked out the road past the No Border Kitchen encampment. As I approached it

I heard shouting. It was explained when I saw some of the young men enthusiastically playing football. I passed the gate with its sign welcoming the visitor. As I carried on out the road with the ramparts of Mitilini castle above me, I met four young men. One of them, a cheery one, said hello and held out his hand. I shook it and the hands of his companions. He asked my name and my country. He introduced me to each of his companions. Emboldened, I asked them their country. He told me that they were all from Pakistan, and I thought I detected a vulnerability in his face and voice when he said it.

I went into a restaurant on the near side of the harbour. I ordered fish, with a Greek salad and wine. The fish were small, their names beyond me; it seemed clear that the way to eat them was head and all. I baulked at the heads, aided by my accomplices, three cats that congregated beside me, until they began to squabble with one another as cats do, until two of the waiters came rushing over and shooed them away. Afterwards I walked out the jetty where there were small boats with outboard motors. A few men were foostering in theirs. I watched one in particular, who came along with his daughter of about ten. He began briskly and with purpose tackling the rope that tied the boat to the jetty. But it went on and on and his daughter became increasingly resignedly impatient, her hand clenched underneath her chin as his brisk work went on interminably. At the end of the jetty there was a group of elderly men in a kind of *cabooche*, who perhaps while away their days of retirement there. Close by me there was a car with its passenger seat inclined backwards, and a man who looked an invalid, his fingers preoccupied with his worry beads. There was another man banging an octopus repeatedly on the ground, while a woman sat by, waiting. I decided I would wait until the man with the girl had set off. Eventually, the man had finished with the rope, but spent a further while trying to start his outboard engine, until finally it fired, and they headed off out on the silkily calm water. It was a beautiful early evening, with still clouds in the blue, the sunlight filling everything in deep luminous colour, and I saw clearly the magic of the Mediterranean.

As I came back I heard the calls of men playing a game amidst old boats that seemed to be in dry dock. They were four refugees, playing volleyball with an improvised net, probably two or three generations of the one family. They were entering enthusiastically in to the game. As I passed them I heard a baby crying, and then, amidst the semi-dereliction of the scene I saw a small camp where the family must live, and I caught a glimpse of the baby's mother, rocking, comforting it. I passed on by, and returned to the church to attend the mass that was to be said by the bishop. I entered the small space and was one of a small congregation of about two dozen. A priest came out who didn't look like a bishop to me. He had a wide mouth, set as if in a sombre view of the passage of life, until he sang in a voice that was surprisingly sweet. He gave a long vigorous sermon, all in Greek as the mass was. When the consecration came it was the Welsh woman, Jenny, who sounded the bell, three times for the bread and three times for the wine. At the end she told me that the priest was not the bishop, that the latter was taken up with meeting envoys from the Vatican, in preparation for the Pope's visit at the end of the week. She told me also that the priest's sermon had been about the refugees, and the necessity to be kind to those who had landed on the shores of Lesbos. I thanked her for her kindness to me.

Back out again along the pier was in darkness. A man came near me, with a torch on his head, positioning himself for fishing. The water was still silkily calm. I thought how it would have been a night much better than the previous one for heading out in a fragile craft. A powerful boat came from the inner harbour and passed out into the dark sea. I couldn't tell what kind of boat it was, but I guessed that it was not fish but human souls they were after. The passage of the boat set off undulations in the water that lasted for a long while after it had gone out of sight. Again there were stars in the clear sky, and a crescent moon. The lights of Mitilini were brighter in their nearness than the lights of Asia.

What filled the hearts of those refugees who looked across at our lights on such a night? With what longing must those thousands of eyes have looked across the narrow straits before embarking on their hazardous journey? Their setting out must have been an act of submission, the distant lights of Europe glimmering in front of them: an uncertain voyage, out there on a small craft in a huge sea, their fate in its hands. I thought of the meagre challenges of my own journey, compared to the hugeness of theirs, and of all the millions who have similarly set out already this millennium. How can we do them justice?

Back at the hotel, as I was going up past the lounge, I saw seated around a few tables put together a group of clerics who must have been the Vatican envoys. They were in the company of the bishop, and I recognised the priest who had said mass earlier. I bought a bottle of beer and drank it out on the balcony. Outside under the crescent moon I thought, at that particular late juncture, there was no place I would rather be than Mitilini harbour.

The next morning before leaving, I loitered for a couple of hours out on the pier, gazing around, in to the harbour and across the water towards the coast of Turkey. There was a boat stationary off the shore that might have been the one of the British navy, and a small fishing boat plying up and down close by it. I sat a long while watching and listening to the water against the pier. Behind it, I could see on the hill the ramparts of the castle. The No Border Kitchen refugee camp was quiet beneath it. It looked idyllic: a quiet encampment, behind the beach. The battlements above it were the only portend in that image of the external world that was about to visit. A couple of weeks later, I saw a report on the internet of the closure of the camp, and the transfer of its residents to the detention centre at Moria. There was phone footage of a resigned procession of young men, being marshalled out of the camp onto a bus by the police.

A bus brought me early back to the airport. Before I went in I crossed the road and sat close to the bus shelter of the first morning. Down on the shore I examined the detritus that was still scattered there: along with the remains of the dinghies I could see a few rubber tubes, an abandoned coat, and what looked like a woman's party dress that will never be worn again.

The flight was hardly more than a half an hour before we were landing in Athens. Back in the Attalos Hotel, I had been given a room out the front with a balcony. I visited the balcony for a little while before heading back to the roof-top bar. I sat with a bottle of beer, at a table with an unimpeded view of the Acropolis. The bottle of beer was followed by another. Two American men behind me were talking about medical work with the refugees. One of them had just arrived in Athens, and the other was briefing him, mentioning Piraeus, Idomeni, and Lesbos. I took them to be doctors working for some NGO. Impressed by their commitment, I would have liked to join their conversation, but my native shyness inhibited me. Then, before they left, they came up close by me, taking photographs of the Parthenon, and I, influenced by solitude and alcohol, asked them what did they think of its significance. One of them mentioned the dawn of Western culture. I, in my inebriation asked might that building on the hill have been the setting for a priestly bureaucracy who, without fellow feeling, ruled the people below heartlessly. I had meant it as a contrast to the heart of their work, but they seemed taken aback by my intrusion, and very speedily they melted away inside into the hotel.

I headed back out the street to Monastiraki. I purchased a souvlaki, admiring the speed with which the small man produced the package that he handed to me. I crossed the road into the square, and did what I had intended to do: I purchased small lights that their sellers were catapulting into the warm Athenian night air. The one from whom I made my purchase said he was from Bangladesh. I told him my memory of buying the same toy in Rome from one of his compatriots, and he and some of his fellows laughed in recognition that that truly would be the case. I thought of the solidarity of the refugee and the economic migrant as I walked back to the hotel with a can of beer I had bought, which I consumed out on the balcony above the loud crowded street, before I came in to the room and closed the door on my final night in Athens.

The next morning, as I checked out, the man behind the desk asked me how were things in Ireland now. He said he didn't believe anything he heard on the news about the economic crisis, either in Greece or anywhere else: that those who have money hold onto their money, regardless. My flight home was not until the evening. I left my bag in the hotel and took the metro to Piraeus. A ferry had just docked. There were groups of Eastern people standing around on the quay. There was a quickened atmosphere about the place. Some police were in attendance, a few with riot shields. A policeman was interviewing two young Asian men who seemed as if they had alighted from the ferry. They had no baggage with them, but I surmised that all they had to smuggle was themselves. A family of refugees passed, and boarded a bus, along with others. A group of Syrian boys and girls walked by. One of the girls was pointing out and laughing at a young Greek couple kissing at the edge of the quay. I came across an encampment of refugees on a grassy island beside the road. I noticed one young family who seemed oblivious to traffic passing around them; they were engaged with one another as if they were in their own private domestic space. The man was entertaining an infant as the woman, doing some domestic task, was smiling over at them as she talked to him. I noticed her perfectly formed white teeth, and her rasping cough.

I was back on the metro to Monastiraki when a woman suddenly emerged from behind me. She had a baby in her arms and she was inarticulately begging. I took her to be Syrian; she had the fluid poise of women from that region: our Near East. Taken by surprise I found myself refusing, as everyone else was doing. She immediately disappeared like a ghost, silently, and immediately I regretted my refusal. I held a small donation in my hand hoping that she would return but she didn't. I looked for her at Monastaraki, but I couldn't find her.

A few hours later I flew home, from the south-eastern periphery of Europe to its far north-western shores. A few days later, Pope Francis was visiting Lesbos. I happened to be back on another island: in the old schoolhouse on Collanmore, in Clew Bay. The school had been open for just seventy years, from 1887 until

1957. The old classroom was now a living room. The voices of the children singing out their lessons were now stilled in the silence of its stones. The nearby ruins of the islanders' houses remained a testament to the desertion of the island, most of whose families had set off as economic migrants, across the Atlantic or the Irish Sea. It was only when I returned to the mainland that I saw images of Pope Francis's visit to Lesbos. He was accompanied by Patriarch Bartholomew of Constantinople, and Archbishop Ieronymous, of Athens. After they visited the camp at Moria, their motorcade made their way to the harbour of Mitilini. Standing at the edge of the quay, Francis made his speech. He expressed his admiration for the generosity of the Greek people, sharing the little they had with those who have lost everything. He said that refugees and migrants, rather than simply a statistic, are people first of all with faces, names and individual stories. He said that it is only through service to others that we can get beyond ourselves. He said that Europe is the homeland of human rights, and that whoever sets foot on European soil ought to sense this, and thus become aware of the duty to respect and defend those rights. He remembered those who set out on the journey, but had foundered, unable to reach the shores of Europe. The ceremony concluded with Francis joining his companions as each of them threw a wreath into the waters that had carried so many, and in which so many had drowned. God rest them.

A DAY IN DUBLIN

I drove in to Sligo Station in the early morning darkness. There were slivers of light as the train headed east. I was looking through the book that I had brought with me. It was a book of photographs of the universe. Sometimes I looked out the window. The colours in the photographs mirrored the colours outside in the brightening sky. In the carriage opposite me there was a man with a book whose title I saw: *Change Your Life in Seven Days*.

In Dublin, when the train stopped at Connolly Station, I walked quickly to Stephen's Green where a Luas train was waiting. It brought me along Harcourt Street, through Charlemont. Coming into Ranelagh, my ticket was checked by a woman from Africa. I saw her chatting afterwards on the platform with a man in a similar uniform, who looked to be Caucasian. The train carried on through Milltown, and it seemed in no time at all that we passed over the new bridge into Dundrum. The precincts of my childhood: it used to be called a village; the main street was now reduced almost to a facade with a by-pass behind it. It put me in mind of a film set. The church seemed the only thing unchanged. I carried on past it to the vast new shopping centre that has appropriated the name, Dundrum Town Centre: a consumerist cathedral where I paid my brief dues. I retreated with relief to the main street of the village. My feet drew me in to the church. Inside, I stood at the back remembering a period past that remains only in the memory of a store of people that is steadily diminishing. The church was almost empty: a few figures, who looked as if they had reached their final furlong, were sitting separately, silently praying. Laconic sounds were echoing. There was a woman spraying the flowers on the altar. Behind her the window had a picture of the crucifixion, and above it a bearded man, who did not now look old to me. He was holding a book or a tablet, with the letters Alpha and Omega. On a side altar, there was a shrine with candles. I was searching for where to put my coin when a woman came up to show me where to put it, in the security of the wall. I lit one candle, for memory. I remembered reluctant adolescent Sundays, standing just inside the side door, then increasingly going no further than the porch, barely going through the motions, before abandoning it in disaffection. A Dundrum childhood was giving way to a complex urban world where the image of the self that was growing up into it was too much constituted by the imagined estimation of others. On my way out, there was an old woman in the porch who was telling another woman of her medical complications. They made their exit. I followed them, dipping my hand in a bowl of holy water. On the way back in to the city, at Milltown the train passed over the Nine Arches. Looking down at the Dropping Well pub in the valley below, I remembered late one night a long time ago, walking along there on my way home, looking up above the dark deserted arches at the moon appearing and disappearing behind the racing clouds. A moment of youthful resolve. A glimpse of authenticity.

Back in the centre I was drawn in to Stephen's Green by the sound, then sight of a few followers of Hare Krishna, padding along with bells and chanting. Two young children with faces painted were walking beside them. The lead man was smiling and enthusiastically greeting the impassive observers on their park benches as he passed them. A group of young men and women gardeners were languidly working in a bed of blood-red flowers. As I walked down Grafton Street and into Westmoreland street, there was a young blind man in front of me, flailing his stick widely, sometimes causing others to jump out of his way. He was giving his full energy to it, marking carefully the many obstacles with his stick. It wasn't until O'Connell Bridge that I approached him and guided him across to the beginning of the central aisle. My gladness about making a gesture of good will was tempered by a doubt about whether he fully appreciated the intrusion on his determined independence. I let him go on ahead of me and waited under the monument to Daniel O'Connell until he had disappeared. I looked up, for the first time giving the monument a proper scrutiny. The figure of O'Connell seemed lost above a circle of allegorical figures. All I could properly see was the bird perched on top of his head. The birds that make no fine distinction between the monuments of the human world.

I walked up the centre of O'Connell Street. The next statue was of the nationalist, William Smith O'Brien. He was a diffident leader of the abortive rebellion that was attempted in 1848. He was convicted of treason, and was sentenced to be hanged drawn and quartered. The sentence was commuted to exile on an island south of Australia, called after a European: Van Diemen's Land. Close by Smith O'Brien's statue was the statue of Sir John Gray, whose plinth proclaims that it was pre-eminently through his exertions that Vartry water supply was introduced to the city. Next was the statue of Jim Larkin across the road from Clery's, with his hands held aloft, and the quote beneath him: "The great appear great because we are on our knees."

The Spire was alone among the monuments on the street in not being the figure of a human with a story, dressed in antique clothes. A monument without memory: it was erected there as a replacement for the pillar that had been topped by the figure of Nelson, until it was blown up by persons unknown, in 1966, before the commemorations that were due to take place, half a century after the Easter Rising of 1916. In the north of the island, the years immediately after 1966 began with civil resistance, before being subsumed in more atavistic forces, and the inauguration of a nightmare that lasted three decades, until an agreement was reached in Belfast as the twentieth century was drawing to a close.

Since it had been erected at the beginning of the new millennium, the Spire had already receded into its familiarity. I saw no one look upward from their quotidian preoccupations. It had become a place of rendezvous like the pillar before it. Unlike in the days of the pillar, the atmosphere was cosmopolitan. In the midst of predominantly foreign faces there was one native Irishwoman, no longer young, who had perhaps taken leave of her senses and was dancing alone in her own world beneath the spire, dressed like someone who might have been going out for the evening to a musical.

I passed on to the monument at the top of the street: the slightly debonair figure of Parnell. On the plinth was a quote from one of his speeches, which did not seem adequate to his political and personal complexity. It began: "No man has a right to fix the boundary to the march of a nation." From there it was a short distance to the GPO. I crossed the street and went back to it. Inside, its ample hall took me blurred moments before I could take it in, surrounded by the windows of the hatches where customers were doing their business. The hubbub. I tried to imagine the hubbub that there must have been during the siege of Easter Week, 1916. I remembered schooldays when that history was proclaimed to us, loudly. The blood sacrifice of the leaders of the rebellion merged in the imagination with stories of the early Christian martyrs, and the child found it hard to distinguish between the political and the religious saints presented to us as moral exemplars. I wondered, if the rebels inside it in Easter Week 1916 had looked out and seen a vision of the street a century later: Nelson's Pillar replaced by the Spire, with faces from five continents gathering beneath it, what would have remained of their nationalistic conviction?

My reverie was interrupted by a man from the Near East, bearing a letter to a government department in his hand, asking in broken English into which box he should post it. I found a gladness in the small generosity of directing him. From there it was with some inevitability that my feet took me into Henry Street. There was a man standing there, holding a pole with a placard advertising a fortune-teller; I heard him tell a woman at a stall close by that the fortune-teller had been in a serious traffic accident, and had died twice, gaining second sight. I passed on into Moore Street. There were two Asian women in a shop there, one of them holding up two pots for the other to choose between them. I bought from a stall some bananas, for the journey home. As I walked back to the railway station, I thought of the liberation of Dublin, brought about by the presence of foreigners, unburdened by our history. I thought of those moments of reprieve when you leave the weight of your own past behind you.

It was dark on the train back. The passing lights outside were fragile terrestrial ones, and for most of the time the windows darkly reflected the carriage back into itself. I thought how, in the life of an individual and of a nation, time unfolds in a chain of unanticipated causation. What has happened inexorably happened, beyond any actor's intention. For us now, as it was a century ago, the world a century into the future is beyond anyone's imagination. I looked again through the book of photographs I had brought with me. There was one in particular, taken in 1968, that has been called Earthrise. It was taken by one of the astronauts in the Apollo 8 mission to the moon. That mission did not land there, but the three astronauts, when their capsule went around the far side of the moon, were the only humans ever to go out of sight of the mother planet. They were also the first humans to take photographs of the earth from a great distance. Then, in their capsule called after the old Greek god Apollo, they chose to read the first verses of the book of Genesis: "In the beginning God created the heaven and the earth, and the earth was without form, and void; and darkness was on the face of the deep ... "

Meanwhile in the West of Ireland, the contemporary quotidian continued to hold sway. One day, I was stuck indoors all day at a seminar in Bundoran. It was led by an expert, but it seemed to me to be a case of the blind leading the blind as I looked out wistfully at the sunlight in the blue, until at last, I escaped, and it was a relief to be out and down the road, through Sligo, Charlestown, Castlebar, Ballinrobe, and into Connemara. I was headed to Rosaveal to catch the ferry to the Aran Islands where I hoped to join a friend who was bringing a boat to Westport. I had been told the day's last ferry was at half six; it was that by the time I reached Maam Cross, but I headed on. Connemara was achingly beautiful on that sunlit evening. There were still pools of light on mountains, on stone walls, on small lakes. I took a wrong turning after Maam, or rather I went straight when I should have turned left. It was only when I reached Patrick Pearse's cottage at Rosmuc that my suspicion was confirmed that I had lost my way. I stopped outside the cottage, at last accepting that the ferry would go without me. It was closed up and deserted, a traditional cottage that served as a refuge for Pearse, before his Fenian zeal took over. I wandered around it and wondered: if he had been given a glimpse of his country one hundred years later, might he have chosen a life of poetry and seclusion? Might he have retreated to live out his life here in this small cottage in the deeps of Connemara, surrounded by the music of Gaeilge, the daylight view out through his window of the shifting light of the lake, and on a clear night a glimpse into the beyond? Would he, (and we), have been better served if he had chosen Rosmuc rather than the GPO?

I retreated back through Connemara, Killary, through Westport without stopping until I reached the tarbert at Cleggan. The sun was setting over Achill, and an almost full moon was rising above Rosmindle as I walked across. The low tide had turned, and I had just enough time to reach the shore of Collanmore as the island was becoming an island once again. It was still bright as I reached the schoolhouse. The miracle of once again returning to it. When I came round the front of the house I disturbed a lamb in the garden enclosure. The fence was down. The bottoms of the fuschia had been eaten away. It seemed a small intrusion, but there was a strong smell of sheep. Then I saw the lamb's mother, head up, caught in netting. There was a stillness beyond sleep about her. She was dead. The lamb returned and was sitting by her, still drawn to its dead mother. They must have been like that all night outside as I lay inside between waking and sleeping, the image of the morning's task of the removal of a body ahead of me. I braced myself for the detachment required for it and set about it early. The lamb ran away as I approached. I pulled the sheep along, still trapped in the netting. It had become an object for me. I pulled it back through the garden, out beyond the wall, and left it upturned there, its lifeless legs spreadeagled. The lamb was lingering close by. I left them there and spent the rest of the morning doing a few chores around the house to set it up for the new season. I emptied out the rain-barrel and washed it in the sea. The tide was falling. When I went back to look again at the sheep, the lamb had left the carcass of its mother, and I couldn't distinguish it among the many sheep and lambs in the field that were seemingly nonchalantly grazing. Later, I climbed the hill. The tide had turned and was once again rising. The boundary between island and mainland was filling with water. Way out beyond the lighthouse, I could see the shape of a big boat coming in the bay. There was a strong sense of a voyage ending.

AT ALGECIRAS

When the plane landed in Malaga I headed out through the concourse to the familiar place where there was a bus waiting. The way in was dark; it was late on an October Sunday night and it was raining. When the bus came in to the centre I disembarked on the Alameda Principal. The streets were almost deserted. I found the Hotel Carlos V, in a side-street between the Alcazaba and the cathedral. The reception was empty until a man came along and checked me in to a room on the third floor. He told me I could get water in the lower lounge. The machine produced two bottles for my euro, instead of one. When the man told me I may as well have the two, I decided to take it as a small piece of good fortune as I made my way up to my small single room. I turned off the light, giving way to the darkness. In the morning the darkness lingered, and I realised that it was a room into which only the faintest of daylight ever enters; outside, the window was half way down a deep well formed by four walls, the sunlight far above. I took the lift back down to the lower lounge and purchased a few plastic cups of coffee from a machine there, to assuage the pangs until I had breakfast in the café directly across the road. Afterwards, I had passed the cathedral when I turned back and paid a visit inside, just to breathe in its echoing presences. Outside again my feet took me inexorably in the direction of the sea. Familiarity resurfaced as I walked along Paseo del Parque, and on to the beach. The morning sunlight was glinting across the water. I followed the example of a few others and went down, taking off my shoes and socks and standing in the shallows of the Mediterranean. My steps brought me back to the harbour. The Pompidou Centre was close by it and, as I approached it, I was startled by the sight of an outdoor exhibition of large photographs of refugees fleeing into Europe, in the immediate aftermath of the voyage of their lives. It is a human flow that has continued across the Mediterranean, from its eastern shores to its western boundary at the strait of Gibraltar. Inside the Pompidou Centre, coincidentally, there was a temporary exhibition with some further representations of refugee journeys: one was in three walls of a room, in which the visitor was surrounded by the sea, to the sound of rowing and glimpses of oars. There was no land visible, and to my eyes it could have been in the waters of Clew Bay. Outside again, with my phone I took some photographs of photographs of refugees.

    I walked out the harbour to a café bar I had visited a few years before. It was unchanged. I had two glasses of beer and a plate of patatas bravas. There was a couple outside at a front table who suddenly seemed to get romantic, and then equally suddenly, the man first, the woman next, each resorted to their mobile phone. The intense look on the young man's face and the urgency of his texting were as unreadable as that of an alien species. I left them to it, and walked leisurely out the quay where I loitered awhile as the day darkened before returning to the city streets. My attention was caught by a sign outside a bakery that there was a café upstairs. It was volubly crowded. I had a dish of spicy vegetables there, admiring the speed and diligence and togetherness of the young couple who were running it. Afterwards I sat outside awhile, as the crowds strolled by in the balmy evening, until I made my way back to the hotel reflecting on the harmony that imbued the place, that imbued me too.

    The next morning, I left the hotel before dawn. There were few people about as I walked to the beginning of the beach. I sat on a bench above it and could hear faintly the sound of the sea as the sounds of traffic began to build up behind me. I went down to the water's edge in the first glimmer of light. It was only when the sky reddened with sunrise, and the sun itself put in a brief appearance, that I noticed a couple of figures who had been sleeping on the beach, secluded by umbrellas and a pile of deck chairs. I imagined them to be migrants.

I walked up the Alameda Principal, turning off in the confidence that, despite the fact that the streets had not surfaced in my mind in the previous few years, I would find the way. The bus station was immediately familiar. Its café had been extended and refurbished. I thought I recognised one of the men serving. Caught up in his quotidian busyness, he showed no sign of recognition as he served me. I boarded the eleven o'clock bus for Algeciras. There was an Arabic ambience to some of my fellow passengers. I spent the journey looking out at the passing landscape with glimpses of the sea, until the sudden apparition of Gibraltar. Shortly afterwards, the sight of chimneys, cranes and containers announced the onset of the port of Algeciras. I passed out of the bus station through a raggle-taggle crowd around the door. I passed a tourist office with a sign that said it was closed for technical reasons. In contrast to Malaga, the streets looked down at heel. I was coming near the square when a young woman sitting in an alleyway called out "amore amore". I wondered how many years since I received such an importunate solicitation. I passed on. The streets soon gave way to the busy port that is one of the main gateways between Europe and Africa. I walked along across the road from the port, following a sign for the Hotel Reina Cristina.

Yeats and his wife, George, landed in Gibraltar in November 1927, and made the short journey across the bay to the Hotel Reina Cristina. He was suffering a severe bout of ill-health, an inaugural challenge of his old age. From the hotel he wrote the following in a letter to his lost muse, Maud Gonne: "My dear Maud, A multitude of white herons are beginning to roost among the dark branches of the trees outside my windows. They fish in the Mediterranean on the other side of Gibraltar which is some ten miles off, and then fly home to the gardens here for a night's sleep."

It seems that Yeats was in a state of great anxiety while they were staying in the hotel. His wife George wrote a letter home to an associate, Lennox Robinson, in which she said: "WB of course is making his last will and testament at all hours of day and night, hurrying to finish a poem but has not been able to begin yet. 'Of course I shall never be able to go on with the autobio now…' etc etc. All poppycock. However, in the same breath he talks of writing a poem on the herons at Algeciras 'in a few years' time.' What a pillaloo."

George's ironic attitude is corroborated by six photographs of the poet, looking relaxed, and perhaps inebriated, in the gardens of the hotel, along with George, and a friend, Jean Hall; unguarded images of a man, most of whose photographs seem to strike a solemn pose. Despite his anxieties, they are photographs of a man who had more than a decade left, of flourishing and productivity.

One of the early fruits of this was a short poem, in the first verse of which the herons of Algeciras were transmuted by Yeats into what he called, "heron-billed birds", that he imagined flying across each evening from their feeding grounds in Morocco to settle in the rich midnight of the garden trees of the hotel. The second verse is set in the Rosses Point of the poet's childhood. He remembers a day he spent gathering shells from Rosses' level shore, and bringing them back in the hope that he might receive the commendation of an older friend. The third and final verse is set at close of day in either place, in which he imagines being questioned by "the Great Questioner": how he might with a fitting confidence reply.

He called the poem: "At Algeciras – A Meditation on Death".

There was an insalubrious aura to the steps I climbed up to the grounds of the hotel, but the entrance was grand. At reception, I was dealt with by a young man. Neither he nor the woman in the background could speak much more English than the dearth of my

Spanish. Beside the reception, I looked at a plaque with autographs of celebrities and luminaries, looking to see the name of W. B. Yeats. When I asked, they hadn't heard of him, but the young man pointed out to me a framed photograph behind us of two men: one of them looked out of the frame with the haunted face of Gabriel Garcia Lorca.

I was given a room much more salubrious than my usual hotel accommodation. In the hotel booklet there was a page that mentioned that the hotel was almost entirely burned down in 1928, presumably obliterating the record of the Yeats's stay there the year before. The room had a view as I had requested overlooking the gardens. I took the opportunity to explore them, and I sat for a long interval around five in the afternoon in a secluded place among the trees, in which there were many song birds singing, but I caught no sight of a white heron. Later, in the dusk light I went out to the gate and crossed the road for the view of the port. A sign there spoke of the heyday of the hotel, when there was a beach below it, down to which the guests would descend on the steps that I had climbed as I arrived; steps that now end in a sudden hiatus at the side of a main road. I loitered on awhile. Above the industrial noise of the port, there were occasional cries of what sounded like gulls, but I caught no sighting of any heron in the trees of the garden. I guessed that the many lights and moving traffic had deflected the birds from their traditional flight paths. One possible heron-like bird appeared briefly before disappearing behind a chimney. Before I went to my room, I wandered the corridors, looking at old photographs of the hotel and its environs, particularly a few of the beach that is no more, and the bay between there and Gibraltar that was open sea then, but for the small island called Verde, that had on it what was called a fort, indistinct in the old photograph. However, the prospect that Yeats and his fellow guests would have had was clear, before the green island was enveloped and engulfed by the port that rendered the hotel an old-world island, marooned, surrounded by the industrial new. I was glad to spend the evening in the hotel, feeling an empathy with its maroonedness in the wider world.

In the morning I took a local bus to La Linea, and walked across the road to the border, where my passport was checked on both sides, cursorily, as I stepped onto Gibraltar. The name has a quintessentially British ring to it, but hidden in it is the name of the first Muslim conqueror to come across the straits: Jebel Tariq, meaning Tariq's mountain. I passed through the streets with their Union Jacks, their shops selling fish and chips and souvenirs of England. The mountain that used to be Tariq's loomed above them. When I reached the square, I boarded the number two bus to the most southerly tip of the peninsula, that is called Europa Point. Most of the passengers on the bus were elderly, and most seemed acquainted with one another, talking in rapid fluent Spanish. The woman beside me was engaged with a man in front, from which I could gather they were talking about the planned exit of Britain from the EU. Suddenly the man switched to English as he said in a seemingly native upper-class English voice about the British government, that they should stop negotiating, that they should get up from the table and just say goodbye, we're leaving. The journey was along narrow twisting roads. Most of my fellow passengers had got out before we reached the terminus. The most prominent building in my sight was a mosque, but I went first to the lighthouse, and the memorial to the Polish General Sikorski, whose aircraft crashed there in 1943. The day was murky and windy. I looked down into the turbulent water. The area around the mosque was a building site with many buildings under construction, one that looked like an office block, or a block of apartments, right next to the mosque. The gate was closed, but it opened when I tried it. I tried one door that was locked, but I found an open one around the back and I went inside. A man appeared and asked me did I wish to visit the mosque itself. He showed me the door, and said I should take off my shoes. Then he turned on the lights and disappeared, and I was alone in the large, spartan spiritual space, standing in my stockinged feet, imbibing the silence. The man reappeared as I was leaving. When I thanked him, he said you're welcome, and made a gesture with his palm pressed briefly against his chest.

Back in Algeciras, I sat awhile in the gardens of the hotel, close to a fountain, whose flowing water masked the noise of the port below. I thought of Yeats, wondering where he had been when the photographs of him had been taken. I brought up on my phone his poem, "At Algeciras", wondering was there still the faint ghost of his presence there. The day was darkening early. I saw no herons, but some big gulls were circling and swooping, some seemingly settling in the trees close by. I thought that no heron would now risk the flight across the bay from Gibraltar, above the industrial lights and noise below.

Next morning I walked down the quay to the terminal building where tickets were sold for ferries to Tangier and Ceuta. I responded to the call of a man in one of the cubicles, who very quickly had sold me a return ticket for Ceuta. Soon, I was heading out on the ferry, past the rock of Gibraltar on the port side, across the narrow strait on water that might have been Mediterranean or Atlantic, or the two of them mingled, until we reached the shore of Africa about an hour later. I marvelled at the ease of it. In Ceuta I marvelled at the ease of wandering up past the surprise of a Lidl supermarket to the centre of the city where I wandered and loitered in the busy streets. To the eye of this brief and distant observer it seemed a harmonious intermingling of southern Europeans and northern Africans, all about their sociable business on a sunny morning. I went in to an Arab market and managed with some gesticulation to purchase three bags of spices to bring home. At the end of our transaction the man seemed a bit bemused by me as he proffered his hand. We shook warmly.

Back out on the street I sat amidst the comings and goings, including a very large cheerful man who was selling lottery tickets, and a woman with presumably her Arab parents, who were meeting for what seemed a special occasion, of which many photographs were being taken. I ambled along the street with frequent breaks for sitting and gawking. I reached a square with a church to Saint Francis: the Plaza de los Reyes. There was a gathering there of about 20 youths who I assumed were migrants, listlessly lounging with pieces of cardboard that they used as mats both for lying on, and for praying. There were many local comings and goings. Some were feeding a large flock of pigeons that seemed to be a habitual presence. A young couple came along with an elaborate camera, and started to film the migrants whose demeanour seemed one of passive acceptance. I left the square, and soon came to a view of the sea. I climbed down through a path of small houses. The sun was entering a late quarter as I sat a long while on a concrete bench, looking at the sea, and across the bay, particularly at a line that I wondered was it the border fence between the Spanish enclave and Morocco. Suddenly, two young men came from behind me. They each had black beards, and their garb was unmistakeably Muslim. One of them deftly leaped over the rail onto a narrow ledge directly above the sea, as if it was a common thing for him to do. It seemed as if they were slightly disconcerted to see me and they moved away momentarily, but they eventually settled on a spot close by me, and they both began praying, prostrating silently. When they were finished they stood a while leaning on the rail above the sea. When I asked them could they speak English, they said they could a little. They confirmed that the line I pointed out to them was indeed the border. One of them said he travelled through it daily on his way to study in Ceuta. He pointed out a hill across the bay, towards which they had been praying, and said that it was his homeplace. We chatted for a little while, both of them speaking respectfully to their elder, and we shook hands as they were leaving. The encounter seemed a gift, out of the blue, like life, and it seemed to exonerate the decision to cross to Ceuta, and the day.

Back at the terminal, as I waited for the boat for the return journey, three men came along, two of them comforting the other, older one. It seemed they had all been drinking. The older man was close to crying. I saw him again on the ferry, his eyes closed as if sleeping. Out again on the choppy water of the strait, we passed two small boats seemingly battling, and I couldn't help but wonder: their vulnerability in the immensity put me in mind of a quote I had read, from one of the migrants who had successfully made the journey across the strait to Europe: "Death happens but once; we prefer to risk our lives than to stay there."

I watched from the deck as Africa quickly receded behind us, and the sun slowly descended into the Atlantic beyond the strait. The rock of Gibraltar gradually reared ahead of us, and it was looming to the starboard as we turned back in to the harbour of Algeciras. On my way back to the hotel, I came across a path that was somewhat secluded from the noise and sights of global capitalism around the port below, and I was able to look down at the business of the port, and see it as separate, and at a remove from the world in which I was walking, with the grounds of the hotel beside me. A few sea birds on a roof had me wondering did they have beaks like herons, and, on the edge of the roof of the hotel, the bird I had seen the previous evening was seemingly nesting beside the chimney. I didn't leave the grounds of the hotel for the rest of the day, feeling at home in the world after crossing to the continent of Africa without leaving Europe. I went down to the old gate through which the guests would go as they descended to the beach that is no more. I sat close to the swimming pool. A high port tower in front of the hotel no longer seemed intrusive to my eyes that had begun to let the surroundings be. I loitered harmoniously, as the sun slowly sank behind the trees, and the garden was shaded, and it was evening. I saw no Yeatsian heron, but there was a myriad of small birds vividly inhabiting their garden world. I marked, against the background of the constant industrial monotone, the subtle nuance of their song.

Coming up to dusk, many big sea-birds, that I assumed were gulls, were chaotically filling the air above the trees, squawking. I wondered were they the birds that Yeats saw, and subsequently distilled to become figments of his imagination. I thought of him: that sometimes smiling public man, a publicly successful life behind him, and thoughts born at Algeciras of the death that awaited him. I lingered outside in the warm night, marvelling the phenomenon of the place. I was walking back in through the terrace when my phone rang with an unknown number. At first the person did not seem to get through. It rang again. When I answered it, it was a voice from home, telling me of the tragic death of a young woman, a beautiful soul, found drowned off Coney Island in Sligo Bay.

Before going inside, I fixed my attention briefly on the one star that was visible to me. I thought: one couldn't say the universe is without meaning as a microcosm of it, the human being, is so powerfully deeply drawn to seek it. Later, in bed the warmth of the night had me waking in the midst of dreams triggered by the out of the blue phone call, with a jumble of anxious images born of the disturbances of my past quotidian; anxiety too about my health, and fear that I might even give up the ghost on this my last night in Algeciras. I was awake for a long while, wondering would I be able to get back to sleep at all; and of course I thought of Yeats's Great Questioner: what, if I were asked, would I have to say?

It was still dark when I left the bed and opened the window of my room to hear the dawn chorus of the birds prevail for a while over the noise of human machines. The first glimmer was coming into the sky above the strait, and dawn was breaking upon those mingled seas.

The day after returning home, I attended the young woman's funeral in Sligo Cathedral. It was full to overflowing. Scattered amidst the young, there was the re-appearance of old familiar faces. Among them was the bishop emeritus who, although suffering from a serious illness, had come out of retirement for the occasion. In the midst of the Mass, the young woman's sister played on a tin whistle a traditional tune, for which many different words have been given over the ages, but there were no words on that day for the pure poignance of her playing. The bishop in his homily embodied more than expressed what was inspiring about his faith. He lauded the size of the congregation, comparing it with gatherings of the earliest Christians and, as he finished, he struck his hand on the lectern to emphasise his plea to God, to bring solace to the family in their grief.

Having followed in Yeats's footsteps to Algeciras, I followed in them to Rosses Point. I parked below the statue of a woman with arms outstretched that is a memorial to those lost at sea. The sun was sinking by the time I was walking the second beach. A dog came running down past me and into the sea. It was a black and white collie, identical to my memory of the dog who used to be my companion on that shore, decades ago. He ran out into the water, splashing for the sake of it, and it was as if he was an emissary, appearing briefly and quietly, reminding me of a younger world. I loitered in the rocks at the end of the beach, remembering again the years of my youth, mulling memories.

Rosses' level shore: how could one speak of it after so long? I was drawn to it partly because it was beyond speech. The sea was quiet. The sky was reddening behind it. Long dark lines of water were surging, breaking into white in places that lengthened until they joined as they approached the shore. I conjured up Yeats's "Algeciras" on my phone, and thought, whether or not one has faith in the Great Questioner, in the end of the day, if one listens, one is faced with a question that has become one's own. What do you have to say?

I walked back beside the small waves softly landing. I remembered a line in a letter that Yeats wrote to Elizabeth Pelham, a few weeks before he died: "Man can embody truth but he cannot know it." I thought: you do not have to believe in God, or gods, to be able to imagine appearing before an ultimate tribunal, and the account you might give in the hope that your life might be redeemed.

Back at the beginning of the beach, I was sitting on the wall when a woman with young children came along. An older child, a dark-skinned girl, came up to her from the beach, saying, "Look. Look", holding out her hand on which were arrayed bright shells from Rosses' level shore.

I climbed up from the beach, and back the way I had come, looking across to Oyster Island and Coney, and the Metal Man beacon in the channel in front of them. A guide for mariners, the Metal Man had been set there in 1821, and was already old in the time of W.B. Yeats; his brother Jack figured it in several of his works; it is at the centre of an early painting that he called "Memory Harbour".

The Metal Man: the man who never looks back towards Coney. His hand is always out, always in the one direction, slightly downward, pointing towards the channel. His face, a face from a child's comic book, simple, but as opaque as the Mona Lisa, set always in the one pronounced expression, from which one couldn't tell whether he was showing the way, or pointing towards hazards hidden beneath the surface of the water. Once again I tried to find a word for it. The only word that came to me was *baffled*. Why should he not be – frozen in his antique clothes, in the middle of the channel between an island and the fast-flowing human world? I waited until his light began to go on and off in bursts of three. On either side, Blackrock lighthouse and Oyster Island lighthouse were flashing softly. An evening chill came on the air. I heard a boat leaving the harbour. I could barely see its helmsman in the twilight as he headed across towards Coney. I could hear the engine reversing as it came in to the harbour there. In the background was the rhythmic rush of the surf on the far side of the island. As I walked back the boat was returning. There were two figures now, their voices raised as they chatted above the engine. I saw them disembarking at the quay, but they had disappeared by the time I passed it on my way up to where I had parked my car. I drove home in the darkness.

AUTHOR'S AFTERWORD

Spring in 2020 was a strange terrible season. It marked the onset of a coronavirus epidemic that spread rapidly across human borders. There were those who saw it as little more than a passing flu; others saw it with much foreboding. Its progress was beyond the grasp of any philosophy. By the time that summer came, it was beginning to look as if it would make a lasting mark on human society, and become perhaps, at its worst, an invisible worm with the capacity to destroy the heart of the human race.

It was the first truly global pandemic. But it was not unprecedented. Plagues and pestilences have visited our species throughout recorded history. One occurred during the reign of Marcus Aurelius in the second century of the Christian era. He made only passing reference to it in his *Meditations*, saying that infection of the mind is a far more dangerous pestilence than any unwholesomeness or disorder in the atmosphere around us; that, insofar as we are animals, the latter attacks our lives; but the former attacks our humanity.

A few centuries later, in the early 540s, there was another outbreak, during the reign of the Emperor Justinian. By then, the empire had become officially Christian, and had moved its capital to Constantinople. It was recorded to have killed about one fifth of the population of the city and spread all around the Mediterranean world. Justinian is recorded as having suffered and survived it himself, when it was at its height, in 542. Five years earlier, by his orders, a new basilica had been inaugurated in Constantinople. It was called the Aghia Sofia.

The twenty-first century pandemic quickly spread throughout Europe, and borders were once again closed as they had been a few years before against the flow of human migration from Africa and Asia ... Country after country went into lockdown. Many were living increasingly virtual lives. The West of Ireland was at least for the moment secluded from the worst effects, but was part of a national lockdown that narrowed down freedom of movement until we were confined to our immediate dwelling places. A quiet descended on the country roads, effulgent, in day after day of crystal clear spring sunlight until, gradually, the lockdown was easing as the glorious spring was giving way to a summer of mixed weather.

On the second of July I got up in the first brightness. As I headed off in my car I listened to the early morning news headlines on the BBC World Service. There was a report that a high court in Turkey was about to make a judgement on the legality of what was then the Museum of the Aya Sofia, having been stripped of its religious status by the secular government in 1934. Thus was initiated a probably predetermined process that came to fruition a mere few weeks later, on 24 July, when the call to prayer once again rang out over the rooftops of Old Istanbul, from what had become once again, the Grand Mosque of Aya Sofia.

I was on my way to the old schoolhouse on Collanmore for the first time since the beginning of the lockdown. There was a feeling of release as I breached the border into Mayo. A morning mist was hovering above some of the fields beside the road as the sun appeared above a low bank of cloud behind me. By the time I was near Westport the sky was clear, but for a ribbon of cloud surrounding the church on top of Croagh Patrick. I drove through the town without stopping until I reached Rosmindle. The sun was shining, and the high tide was turning as I pulled the boat in. I let the flow carry me out the mouth before I started the engine and I motored across to the island. The crossing was calm and was easy. I pulled the boat out on the island mooring and walked on up from the shore.

The schoolhouse was abiding, its palpable presence silent and still. Inside, I made myself some tea and sat with it at the desk in the living room, thankful to have made the return. I had brought a draft of the foregoing collection of essays, to give them some critical scrutiny. I placed it on the desk and opened it, but I didn't get beyond the first page; I was stopped up almost immediately by the second paragraph, about the visit to the Aya Sofia. If the author of those words had been writing them this present day, he would have written differently. He had not considered then the possibility that the status quo of the Aya Sofia as a museum might turn out to be a brief anachronistic secular interlude in its long religious history.

I sat awhile perplexed, considering whether to change it, but eventually I left the draft untouched on the desk and headed out into the island day. I walked up the field at the back of the schoolhouse until I gained a view of the island of Clynish behind. There were big round bales scattered around a shorn field beside the house. A fast boat came out with one of the Gavins at the helm, across through the tarbert and in to Rosmoney quay. I carried on walking out the brow of the headland. Islandmore came in to view across the channel, its small handful of houses clustered in front of the shore, a half century since we first landed there. I remembered our visits to the kitchen of the Gills, who are all long gone now, their house a ruin. God rest them. Behind Islandmore, the village of Mulranney nestled on the northern shores, on the road out to mainland Achill Island, that I could see rising at the north-western perimeter of the bay.

I climbed the hill. The view from there was expansive and wide and blue, and deeply familiar. I lingered looking all around me, drinking it in. (That place of islands. My life measured in it.) Beneath me was Collanbeg. There was a boat moored at the jetty, but the door of the house of Joe Jeffers was closed and it seemed deserted. A big boat passed in front of Inishlyre, out past the lighthouse on Inishgort, into the wider waters, presumably on its way to Clare Island. I could see its village and the hill above it glistening at the entrance to the bay.

Inishturk was distinct in the distance to the southwest. Gradually, I brought my gaze back in along the southern shore of the bay, past Old Head and Louisburgh, until I was taking in the dark presence of Croagh Patrick, looming. That place of pilgrimage. As I looked, the ribbon of cloud cleared and I could see the white church at its summit, from where the pilgrim, when the challenge of the climb is completed, can look down and survey the whole wide world of the bay. As I turned back through the fields to the schoolhouse I thought: there's no road; once you cross to the island, there's nowhere else to go. You're there.

ACKNOWLEDGEMENTS

These eight essays were written over the course of more than a decade. The fact that they have all been published in one place, by the good offices of *Irish Pages*, has allowed me to see the continuity between them, and to hope they might be seen by the reader to form a unity.

Thanks to Manus, for the long-lived Conversation.

Gerard McCarthy, 2020